Astronomy in Popular Culture

To my step-grandson Teddy – a stargazer in the making
Brian

For my parents, my sister Caroline and soul mate, David
Katrin

Astronomy in Popular Culture

Brian Jones
Katrin Raynor

WHITE OWL
AN IMPRINT OF PEN & SWORD BOOKS LTD.
YORKSHIRE – PHILADELPHIA

First published in Great Britain in 2026 by
Pen & Sword Military
An imprint of Pen & Sword Books Limited
Yorkshire – Philadelphia

ISBN 978 1 39903 655 9

Typeset by Mac Style
Printed and bound in India by Replika Press Pvt. Ltd.

The Publisher's authorised representative in the EU for product
safety is Authorised Rep Compliance Ltd., Ground Floor,
71 Lower Baggot Street, Dublin D02 P593, Ireland.
www.arccompliance.com

For a complete list of Pen & Sword titles please contact

PEN & SWORD BOOKS LIMITED
47 Church Street, Barnsley, South Yorkshire, S70 2AS, England
E-mail: enquiries@pen-and-sword.co.uk
Website: www.pen-and-sword.co.uk
or
PEN AND SWORD BOOKS
1950 Lawrence Road, Havertown, PA 19083, USA
E-mail: uspen-and-sword@casematepublishers.com
Website: www.penandswordbooks.com

Contents

Editors' Foreword

The book you are holding is not what would be considered a 'guide to astronomy'. Rather it is a non-technical guide as to how our view of astronomy and the universe at large have managed to enter our everyday human culture and experience. *Astronomy in Popular Culture* is aimed not only at the beginner or near-beginner, but also at seasoned astronomers and enthusiasts who are perhaps experienced in the subject, but are unaware of the way that references to the wonders that the universe holds manifest themselves in everyday life.

Within the pages of *Astronomy in Popular Culture* you will find a rich blend of subject matter, covering areas that you may not have thought of and which are introduced below and throughout the book. They range from the small screen to the big screen, from books and mass media to art and astrology and from the calendar to navigation. Astrology, music, the wonder of the Full Moon, animal navigation, the cultural significance of the Space Age and even a revealing survey of our planet by interstellar visitors also feature, as does the increasingly frequent and predominantly media-fuelled mispronunciation of two of the most-used names in astronomy.

Astronomy and space exploration now feature regularly on our screens and factual astronomy programmes are incredibly popular too. As a result, more people than ever before are more aware of our place in the universe and the importance and benefits of not just astronomy but science in general. The article *Astronomy in Film and Television* by Stuart Atkinson examines how much impact they have had.

Outer space is also the setting for many sci-fi shows, but not all are accurate and realistic. The aliens and alien worlds shown in *Doctor Who* and the various shows of the *Star Trek* franchise are fascinating and exciting, but they give their viewers false and unrealistic expectations about what is actually 'out there', just as Hollywood's sci-fi do. But while some astronomers believe that these hugely entertaining stories risk 'dumbing down' their audiences with their bad science and impossible scenarios, others think that we should not take them so seriously, and enjoy them as pure entertainment. We will be looking at both sides of the argument.

What would extraterrestrial visitors make of Earth? The article *Report on Planet Three* by Philip Wallace looks at the way visitors from elsewhere in the

Galaxy may analyse Earth, their thoughts and comments being presented in the form of an extraterrestrial's report on our planet and the environmental stewardship of its human population. The alien visitors look at the Earth, assessing it as a biological system as well as a chemical and physical one. The climate is seen to be a principal concern due to the fact that the Earth's atmospheric envelope is our interface to space, and the key to the planet's wide scale habitability.

Astrology has been part of human culture for thousands of years, and an alarming proportion of people today still profess to believe in it to varying degrees – from those who just casually read 'their stars' in a newspaper to those who actually allow it to influence decisions in their lives, and even in business and politics. As Neil Haggath tells us in his article *It's Not in the Stars: Why Astrology has no Place in the Modern World*, this is a hangover from an ancient geocentric world view, which has been obsolete since Galileo first turned a telescope to the skies. The article discusses why this archaic superstition should have no place in the modern world.

Since the dawn of Man, the stars and planets were just mysterious lights in the sky, some fixed, some moving. The Sun and Moon were gods, to be worshipped. Then in 1608 a Dutchman, Hans Lippershey, turned the first telescope on the skies. A year later Galileo Galilei did the same, with better results, and it was realised that there are other worlds out there – perhaps with life on them? By 1865 Jules Verne had written *From the Earth to the Moon* a fanciful story in which his heroes were fired from a canon, while in 1900 H.G.Wells used anti-gravity thanks to a material called 'Cavorite'. The idea of travel to other worlds had entered the public mind, to be realised in 1969 by Apollo. As we see in *Astronomy and Science Fiction* by David A. Hardy, the future is wide open!

Astronomy is inextricably linked to the lives of billions of people through religious festivals and folk traditions that are governed by the cycles of the Sun and Moon: Ramadan, Chinese New Year, Diwali, Easter and many more. *Astronomy and the Calendar* by David Harper explores the deep connections between astronomy, calendars and human culture.

Successful human travelling involves many complications – satnav, compasses, maps and road signs to name a few. We don't find them easy but, as we discover in *Astronomy and Animal Navigation* by John McCue, the arctic tern migrates from the Arctic to the Antarctic and back again every year – a round trip of nearly 20,000 miles, the longest trip of any bird. How does it do it? The Painted Lady and Monarch butterflies also famously fly thousands of miles every year, but humpback whales navigate underwater. How? Even dung beetles and seals can find their way by the stars.

Before cameras were invented, astronomy sketching was an essential tool used by early astronomers when recording their observations. However, for centuries astronomy and space have provided a huge source of inspiration for many artists, such as Giotto di Bondone, Giovanni di Paolo, Vincent van Gogh and Georges Méliès to name a few. As we see in the article *Starry Starry Night: A History of Astronomy Sketching and Art* by Mary McIntyre, this brought astronomy to the forefront of popular culture, where it continues to inspire artists today. This chapter takes a closer look at how the space art genre evolved.

Human beings have used the movement of the Sun and stars in the sky to gauge the time of day or night. Astronomy is intimately connected to time, both through clocks, with their hands that mimic the movement of the Sun, and at the deepest level of physics, where Einstein's theories of relativity show how the passage of time is warped in extreme settings. In his article *Time, Space and Stars: How Celestial Clocks Connect Us to the Universe*, Robert Massey brings that connection to life, from the earliest Neolithic monuments to sundials, astrolabes, the Dutch clocks and orreries that show the positions of the planets and the phase of the Moon, to the atomic clocks that are more accurate timepieces than the Earth itself. It will end with a look at how understanding the warping of time is essential for modern life, from communications to the satnavs in our cars.

As Marc Read informs us in *Astronomy in Everyday Items*, astronomical words and designs are everywhere in the world of brand-names and logos. From chocolate bars to chain-stores, car manufacturers to cigarette cards, even coins and postage stamps, space has been used to suggest the exotic and the futuristic. This article examines the stories behind these ubiquitous elements of popular culture.

In *Astronomy in the Mass Media*, David M. Harland reminds us that astronomy is the most popular of the physical sciences. This is reflected in the coverage it receives by the media. News editors are attracted to its broad appeal, which ranges from the Great Moon Hoax of 1835 to the debate about life on Mars, the ever increasing total of planets known to be orbiting other stars, and the discovery of gravitational waves. How is the subject treated by different news outlets and types of media, and how has this treatment changed over the years, especially since the arrival of the internet with its unprecedented speed, reach, and its ability to link scientists directly to the public? How do the media choose the topics they will report, and on whom do they chose to comment? How have traditional media and science organisations adapted to the fact that people can keep up to date with developments without waiting for the media to report?

In his article *Music and the Cosmos – A Symbiotic Relationship*, Jonathan Powell delves into the work of scholars and musicians alike that has studied and waxed lyrical about the existence of a harmonious link between music and the heavens. Historical records and documented written word from the Greeks muse of not just cultural connections with the stars, but a deeper, more meaningful association whereby the universe itself is alive with its own musical repertoire. The intuitive insight of great minds such as Johannes Kepler imagined the proposed cosmological harmonies that he visualized in planetary motions. From classical to rock, our own musical portrait of the universe has seen rhythmic labours of love depict as with Gustav Holst's 'Planet Suite', a timeless and magical interpretation of the individual characteristics that our planetary neighbours seem to harbour. The celestial tribute continues through the genius of Bowie's 'Space Oddities', to the modern ethereal endeavours of Coldplay's album 'Music of the Spheres'. The intimate interlace threads its way through the centuries and beyond those first musical recordings taken into space on the Apollo missions and the first music track broadcast to Earth from Mars. Humankind's journey into space has gone where the music takes us, and beyond.

As confirmed in the article *Setting Sail on New Seas* by Peter Rea, prior to 4 October 1957 no object from Earth had ever gone into orbit, although some rocket stages had gone into space on sub orbital trajectories. This ability to orbit the Earth opened up new possibilities that would change and assist all our lives. We can now pick up a telephone and route a call to the other side of the world in a matter of seconds. All thanks to communications satellites positioned above the equator. Our weather forecasts are much more accurate thanks to weather satellites continually monitoring the weather patterns and satellite navigation is now so common that we even have them in our smart phones; modern cars actually have them built in. In this article Peter Rea deals with how the space age has impacted on our culture.

The Full Moon is dazzlingly bright, ruling the night sky. Cultures around the world have used it to designate important festivals or simply mark the passing of time, with names like 'When the Snow Blows Like Spirits in the Wind Moon' of winter and the 'Lotus Moon' of summer richly evoking the changing of the seasons. Terms such as 'Blue Moon' and 'Blood Moon' have passed into the vernacular. In her article *Full Moon Names: A Full Moon by Any Other Name Would Shine as Bright* Lynne Marie Stockman investigates this celebration of the Full Moon.

One of the most famous celestial objects of them all – Halley's Comet – has often been the source of confusion when it comes to pronouncing the name. In *Halley: A Question of Pronunciation* we examine how examining the contemporary written record has helped in determining the correct method of pronouncing this iconic surname.

Forming a (perhaps irresistible) follow-up to the above (Halley) article is *Uranus: What's In a Name?* in which the writers make it abundantly clear that, although there are several ways in which the name of the planet Uranus can be pronounced, by far the most popular of these is that which has long been a source of delight to schoolchildren. This article explores the way that the name should be pronounced and details how (and why) alternative pronunciations are often (many would say erroneously) used to try and avoid embarrassment.

The final article *Astronomy Enters Popular Culture* is something of a summing up, where we discover how astronomy has become a small but interesting part of our experiences. The article highlights many different ways in which the public at large now have the opportunity to check the Universe out for themselves, events such as solar eclipses, celestial navigation, sidewalk astronomy and star parties, meteor watching and aurora expeditions all featuring here.

The book rounds off with *Our Contributors*, which contains brief background details of the numerous writers who have contributed articles for *Astronomy in Popular Culture*.

The efforts of a number of people have contributed towards the production of this book, particularly worthy of mention being Mat Blurton – who has done an excellent job typesetting the book – and Garfield Blackmore, David M. Harland, Roger Laishley and Peter Rea who have provided valuable assistance with a number of the images. Our grateful thanks are also due to Jonathan Powell who, aside from his article described above, has provided valuable and interesting input for *Uranus: What's In a Name?*

Thanks are also due to Jonathan Wright, Charlotte Mitchell, Olivia Camozzi, Lori Jones, Janet Brookes, Paul Wilkinson, Charlie Simpson and Rosie Crofts of Pen & Sword Books Ltd for their combined endeavours in the producing and promoting *Astronomy in Popular Culture*.

Brian Jones
Bradford
West Riding of Yorkshire

Katrin Raynor
Pontypridd
Glamorgan

October 2025

Astronomy in Film and Television

Stuart Atkinson

Today's biggest movie blockbusters are science fiction films – fact. Love stories are still popular, gritty kitchen sink dramas still pull the crowds and attract critical acclaim, and the thrilling *James Bond*, *Mission Impossible* and *Marvel* franchises continue to put huge numbers of bums on seats, even if those bums are soon numb because of those films' sometimes excessively long runtimes. But ever since the birth of cinema science fiction films have proved hugely popular with audiences, and there seems to be no decrease in demand for films set 'out there', either close to home, in our own troubled corner of the universe, on some exotic alien world or in a galaxy far, far away.

The same is true of television. Sci-fi has always been popular on the small screen, and many astronomers grew up enjoying – and were inspired by – classic US shows such as *Star Trek*, *Battlestar Galactica* and *Buck Rogers*. The

The Federation starship *USS Enterprise* is one of the most recognisable and beloved science fiction spacecraft ever designed. This model of the original *Enterprise* on display in the Science Museum in London was as popular an exhibit as the real *Apollo* spacecraft and hardware also on display there. (Adobe Stock/David Bond/Wirestock Creators)

Like the characters in all major Sci-fi franchises, the crew of the *USS Enterprise* have been immortalised as action figures, allowing fans to create their own stories. (Adobe Stock/ Willrow Hood)

UK produced its own classic shows too, and although their budgets were not quite as big as their American counterparts, and their special effects weren't as sophisticated, shall we say, people still loved the wobbly sets, cardboard aliens and quarries featured in *Blake's 7*, *Space 1999* and *Doctor Who*. In recent years *Star Trek*, *Battlestar Galactica* and *Doctor Who* have all been rebooted with huge success, and new shows like *The Expanse* have taken Sci-fi TV to new heights of drama, effects and writing.

Of all the above, perhaps *Doctor Who* has had the biggest transformation. Each new

The Doctor's blue TARDIS is now a British cultural icon, up there with Big Ben, red London buses and the Union Jack. Just as they recognise the USS Enterprise from Star Trek and the Millenium Falcon from Star Wars, people who are not necessarily fans of science fiction are aware of how popular the TARDIS is and can't help smiling when they see it. The TARDIS seen here was a bespoke build for a company in Nottingham. Once it had outlived its usefulness for the company in question, it was gifted to the Tony Jordan of the Doctor Who Appreciation Society. (Tony Jordan)

Doctor and companion is unveiled to the world with great drama and fanfare, and after many years of being made for peanuts, a much bigger budget – and much more enthusiastic support from BBC bosses – has meant the show's producers can now afford to tell much more epic stories, with genuinely stunning special effects and guest appearances by many famous and beloved actors, all of which have given the show both a new lease of life and a new respectability.

Many modern Sci-fi movies and TV shows go to great lengths to 'get the science right'. The team behind the American TV series *The Expanse* took great pride in making their spacecraft fly and fight as realistically as possible. But sometimes the science is anything but right, and then astronomers sitting in the dark in a cinema, slurping their fizzy drink and crunching their extortionately-priced popcorn, have a choice: they can either walk out in disgust, or just sit there and be entertained. Personally I am quite happy to take off my science head, put it under my seat and just enjoy a science fiction film for what it is. If hard science is what I want, I will find a documentary to watch – but I know this is something many astronomers struggle with.

Every hero and heroine needs an arch-villain to fight, and The Doctor has fought the despicable Daleks across time and space ever since he first stole his TARDIS all those decades ago. These psychopathic pepper pots, along with their shrill cries of "Exterminate!" have terrified generations of viewers, and although their design has evolved over the years their mission has remained the same – kill everything that is not a Dalek. The example pictured here was built by one of the attendees at The Capitol Six Decades[1] convention, which was organised by Tony Jordan – seen here accompanying the Dalek – Co-ordinator of the Doctor Who Appreciation Society (DWAS)[2] and which took place in April 2023 at the Crowne Plaza, London-Gatwick. (Tony Jordan)

---

1. Located in the fictional constellation of Kasterborous, the Earth-like planet Gallifrey is home to the Time Lords. The seat of Time Lord Government is the Capitol, from which The Capitol DWAS convention derives its name, and which is located in the Gallifrey capital city the Citadel.
2. More information about the Doctor Who Appreciation Society can be found at **dwasonline.co.uk** where there are links to the various activities of the society, including The Capitol convention.

The English science fiction writer Arthur C. Clarke (1917–2008), whose short story 'The Sentinel' – written in 1948 and first published in 1951 – helped inspire the epic science fiction film *2001: A Space Odyssey*. (Wikimedia Commons/ITU Pictures/Creative Commons Attribution 2.0 Generic License)

So, here is a purely personal look at how some of my favourite – and least favourite – science fiction movies have handled astronomy.

Let's start with a movie that got the astronomy right … mostly …

Contact

Although it hardly set the box office alight, for me, the 1997 film adaptation of Carl Sagan's hugely popular novel *Contact* is one of the most perfect Sci-fi films ever made. The story of how a doggedly stubborn but passionate radio astronomer, Ellie Arroway, searches for, detects and then flies across the galaxy to make contact with an extraterrestrial intelligence is handled with great warmth and sensitivity, and features great science too. OK, so the dramatic detection of the grinding WHUMPF! WHUMPF! alien signal is a little over the top, but the film opens with a stunning 'zoom out' sequence from Earth, taking the viewer on a tour of the solar system, accompanied by music from various eras, which still looks impressive, and is astronomically accurate, all these years later. Brief visits to faraway alien stars are stunningly beautiful tool. But for me a scene at the very start of the film is the most accurate astronomically, when young Ellie and her father are outside on their balcony watching a meteor shower. Most films depict meteor showers very inaccurately, showing a sky full of shooting stars, great flurries of them. Contact's meteor shower is much more subtle, with gaps between shooting stars. If you've ever stood in your garden or a farm gateway watching a real meteor shower, stamping your feet against the cold during the long spells where nothing happens, you will know which portrayal is more accurate.

The Dune Films

Since the original novel was published in August 1965, the Dune 'universe' has expanded dramatically. Twenty three books have now been published, including the six original novels written by Frank Herbert and 17 more written by his son and Sci-Fi author Kevin J. Anderson. (Adobe Stock/Cristi)

David Lynch's brilliantly melodramatic 1984 version of Frank Herbert's far future tale of warring political houses and giant worms – which most people remember for Sting's very uncomfortable-looking winged metal underpants – was as cheesy as it was epic, although the two modern films, *Dune* and

One of the most important characters in the Dune books is the planet Arrakis itself. This desolate, bone-dry planet is the only world in the universe where the precious 'spice' can be found. Without it, interstellar travel would be impossible. (Adobe Stock/piai)

Dune 2, are much more serious affairs. When it was released in 2021, *Dune* was the biggest science fiction movie for many years, and although Director Denis Villeneuve took two and a half hours to tell only the first half of the first book of the series, fans loved it. That first film is ponderously slow but looks gorgeous from the opening frame, especially the scenes set in space and showing planets and space hardware. Unfortunately every amateur astronomer and space nut who watched the film pointed at the screen and said "That's Mars! They just copied and pasted Mars into the sky!" during the scene where Arrakis' moon Krelln is shown in the sky, because Mars' huge canyon, Valles Marineris, is clearly visible on the moon's surface. *Dune 2*, released in March

Every year, tens if not hundreds of thousands of tourists travel to the Grand Canyon in Arizona, to stand on its edge and marvel at its grandeur. But compared to Valles Marineris (Mariner Valley) – often referred to as the Grand Canyon of Mars – the Grand Canyon is just a crack in the pavement. Over 4,000 kilometres long, over 200 kilometres wide and up to 7 kilometres deep, Valles Marineris is the *genuine* Grand Canyon of the solar system. (NASA/JPL-Caltech)

2024, has taken the world by storm. With a bigger budget, more exciting scenes and a stellar cast it is a thrilling rollercoaster ride that has had even the most jaded film critics heaping praise upon it. *Dune 3*, due for release in a few years, will have to be something spectacular indeed to beat it.

Stargate

… and talking about cut-and-paste Moons, take a bow *Stargate*, the 1994 super blockbuster produced by Roland Emerich which features a squad of crack special forces soldiers travelling across the galaxy, via a wormhole 'stargate', to take on a deranged interstellar god with a huge tadpole in his stomach. It's a great adventure, just entertaining from start to finish, but after the very 2001-esque rollercoaster plunge through the wormhole the big reveal of the surface of the alien, desert planet of Abydos was ruined for every amateur and professional astronomer watching when the camera panned out to show not one, not two but three moons in its sky – every one of them an exact copy of our own Moon!

2001: A Space Odyssey

Every astronomer will agree that, with its memorable orbital docking sequence set to Strauss' beautiful Blue Danube, and its depiction of the tedium of long duration space flight, Stanley Kubrick's *2001: A Space Odyssey* set the benchmark for scientific accuracy before there was even a single boot print crumped into the lunar dust. But it didn't get everything right.

In the sequences where the lunar shuttle is flying to the site of the Monolith it passes through and over dramatically jagged mountain ranges, with their rocky peaks looking like broken teeth. In fact,

One of the main characters in 2001 isn't even human, it's a tall black monolith, smooth and featureless, created and left buried beneath the Moon's surface by aliens unknown millions or even billions of years before being discovered. When the monolith is unearthed, it sets humanity on a quest for knowledge that will take it to Jupiter and far, far beyond … (Adobe Stock/rolffimages)

as *Apollo* images showed, the lunar mountains have largely been smoothed and worn down by billions of years of micrometeorites and look more like rounded hills than Mount Everest.

Moonraker

With its spindly space station, pew-pew laser beam space marines and magnolia-painted shuttle orbiters zooming around in Earth orbit with their payload bay doors closed, this 1979 James Bond film was made to cash in quickly on the post Star Wars sci-fi movie rush. The film is full of so many scientific inaccuracies they could fill a chapter of this book by themselves, although one stands out. The scene at the very start of the film, where a Drax Industries space shuttle orbiter launches from the back of its 747 carrier plane, destroying it in the process, almost got me thrown out of my local cinema when I shouted, "Oh, come on!" due to the fact that this could not happen; shuttle orbiter main engines drew their fuel from giant external tanks, and only fired during lift-off, so such a mid-air ignition would be totally impossible.

Deep Impact

Most of this sentimental, end-of-the-world-avoided 1998 film is reasonably accurate, and the comet fragment impacts – deep and shallow – are shown quite realistically. The main let down – for me, and for many astronomers – was a line spoken to camera by Tea Leoni's character in the tense run-up to the Messiah spacecraft's landing. "They will fly down the comet's tail … or coma …" she says, explaining what will happen to the watching billions. The problem is that this is totally wrong, the tail and coma being two totally different parts of a comet's structure.

The Empire Strikes Back

I know what many of you are thinking: Oh come on, it's Star Wars! Talking robots! Faster than light travel! It's not meant to be accurate! It's a fantasy! Yes, that's true, but the stirring scene at the end of this 1980 sequel to *Star Wars*, where Luke and Leia watch the Millenium Falcon flying back towards the galaxy, piloted by Lando Calrissian, at the start of a quest to find captured Han Solo, is as inaccurate as it is beautiful. It shows the galaxy visibly turning, which is of course totally impossible, as for such motion to be visible the galaxy would have to be spinning so many times faster than the speed of light that it would rip itself apart.

This image was taken at the Joker Squad UK[3] Legoland Star Wars weekend held in May 2018. It shows members of the society proudly wearing themed costumes depicting a wide range of popular characters from the universe of Star Wars. Donning costumes like this is not unique to fans of Star Wars. "Cosplayers" celebrate their love for many other franchises too, and many travel great distances to attend conventions and fairs dressed as their favourite characters from Star Trek, Battlestar Galactica, Doctor Who and other series. (Paul Mellin/Joker Squad UK)

Also in *The Empire Strikes Back* is a scene which has been copied and shown into so many other science fiction films and TV shows that I've lost count. The Millenium Falcon is shown flying through an asteroid field, ducking and diving, sweeping and swooping to avoid smashing into one of the thousands of tumbling rocks. Scenes like this are a cliché in Sci-fi now, and have led to the public to believe that our own solar system's asteroid belt is just as crowded, with huge boulders tumbling past each other, banging into each other and sending debris in all directions. In reality, the asteroid belt is so sparsely populated that if you stood on one of its asteroids even your nearest neighbour would be so far away you wouldn't be able to see it, and you could fly through it without coming closer than many thousands of kilometres away from a chunk of rock.

---

3. Joker Squad UK was founded in Feb 2013 and currently has over 100 members located across the United Kingdom. The society raises money for charity all year round, their motto being, 'Fun, Charity and Star Wars'. More information about Joker Squad UK and their activities can be found on their website at **jokersquad.co.uk**

The Martian

The *Sojourner* rover, photographed by the Pathfinder lander. Many people who saw The Martian were unaware that this was actually a real rover that trundled across Mars, studying rocks. (NASA/JPL/ University of Arizona)

There are many scientific inaccuracies in this 2015 film (many more than are in the book it is based on), but it would be churlish to list them all because it's such a great romp. But I think it's only fair to point out two. Firstly, the area where *Pathfinder* is dug up – a kind of flat, dusty valley surrounded by high, jagged mountains – looks absolutely nothing like its actual landing site, Ares Valles, which was an ancient flood channel absolutely covered with rocks and boulders, with a few low, rounded hills on its skyline. To be fair, the book describes the *Pathfinder* landing site perfectly, so this was an artistic decision on the part of the film makers. But the most famously wrong thing in *The Martian* is the 'storm' at the start, which strands Mark Whatney on Mars after his crewmates abandon him, believing him to be dead. This is – as writer Andy Weir freely admits – ridiculously over the top. Storms on Mars are slow moving and comprise vast clouds of smothering, choking dust, and could be dangerous to astronauts on the surface, but because they would wreck communications, cover solar panels and ruin seals, not because they are hurricanes which fill the air with flying rocks and boulders. But I happily forgive the film for that 'inaccuracy' for the scene where *Sojourner*, the microwave oven sized rover that the American spacecraft *Mars Pathfinder* took to Mars, follows Whatney around the *Hab* like a loyal puppy.

Armageddon

Back to 1998, and the Sci-fi movie astronomers love to hate, but they shouldn't because it's not trying to be accurate. It is not a documentary; it is a space pantomime – pure, unashamed entertainment from start to finish. If you really wanted to tut-tut at it disapprovingly then fine, you would never launch two space shuttles simultaneously – no matter how cool it looked! – and the film's depiction of the surface of an asteroid as a nightmarish landscape of shining stone shot through with metal spikes and spires is nothing like the images space probes have sent us back from real asteroids, showing them to be dark and covered with rocks like the floor of a quarry. So, while *Armageddon* is about as scientifically accurate as *Button Moon*, it's fun. If you want 100% scientific accuracy then put on the DVD of *First Man* and nod off after half an hour …

Europa Report

Released in 2013, *Europa Report* is one of those 'cult' low budget science fiction films that very few people outside of the Sci-fi community – and even fewer astronomers – have ever heard about, which is a shame because it is actually one of the most accurate and realistic Sci-fi films made for many years. The gritty story of a privately-funded crewed expedition to Jupiter's moon Europa to look for life, the film's scenes set onboard the small spacecraft that descends to the moon's icy surface are extremely claustrophobic, and reflect much more accurately the experiences astronauts will have on the long duration missions of the future than most other films since 2001. But the scenes set outside the lander, out on the fracture criss-crossed plains of Europa, are the most impressive, and the film's Blair Witch-type camerawork really brings the moon to life. When the aliens are eventually encountered they are only seen in tantalising, frustrating glimpses, which makes the film even more fascinating.

And finally, a dishonourable mention must go to …

Moonfall

Astronomers have to put up with so countering so many wild and wacky conspiracy theories these days that it's a wonder they ever have time to swing a telescope at a star. Social media groans under the weight of posts from tin foil hat wearers insisting that the whole *Apollo* program was faked and Neil Armstrong and Buzz Aldrin filmed their moonwalk in a top secret TV studio. Others claim the Mars rovers are actually trundling around on a remote Alaskan island. And then there are the Flat Earthers, still hanging in there after all these years – actually, centuries – of proof that they are talking utter nonsense. But one of the most bizarre theories is that the Moon is hollow, and actually an alien spaceship …

This conspiracy theory is the basis for a film which is so inaccurate scientifically it makes *Armageddon* seem like 2001. When a hollow Moon built by aliens breaks out of orbit and falls towards Earth, only a disgraced ex-astronaut and a conspiracy theorist can save the world, by stealing a retired space shuttle orbiter from a museum and flying it to the falling Moon. Seriously, that's the plot. No, seriously, it is. *Moonfall* is so relentlessly inaccurate that after watching it you want to wash your brain in bleach …

So there you have it, a look at how astronomy has been handled in some of the best, and worst, science fiction films of recent years. There are many others I could have written about of course, some painstakingly accurate, like *Interstellar*, some so inaccurate they are little short of cartoons, like *Gravity*. If I missed out your favourite I apologise, but hopefully I will have given you ideas for some new films to track down and watch.

The USS Enterprise doing what she has always done best – streaking through the stars, seeking out strange new worlds, and new civilisations, and boldly going where no-one has gone before … (Adobe Stock/Willrow Hood)

If you find you want to know more about any of the films or TV series featured in this article, you'll be able to find pages, groups and appreciation societies dedicated to many of them on social media. Run by fans and enthusiasts, these are great sources of exclusive illustrations and artwork, and huge amounts of information about their stories, actors and histories.

At the end of the day, I think it's best – and safer for your mental health – not to take the science in science fiction films too seriously. They are just meant to be entertaining, not educational, and if you spot a mistake talking about it is a great way of educating your friends about the actual science, as long as you are not too aggressive or snobby about it. We do not have, and will probably never build, the warp engines which propel the *USS Enterprise*, but Sci-fi films and TV shows can take us across the galaxy, to strange new worlds, using our imaginations …

… if we allow them to …

Astronomy and the Calendar

David Harper

The passing of the seasons affects all life on Earth. Our distant ancestors looked to the annual return of life-giving rains and the warmth and plenty of summer. They had little need to plan further than a few months ahead, and they took their cues from nature itself, including the stars, whose rising and setting followed a yearly pattern that they would easily have discerned.

Humans began to settle and practice agriculture around 12,000 years ago. A surplus of food eventually allowed the first civilisations to appear in many parts of the world. People began to live in cities with stratified social structures which included many roles that made no direct contribution to food production. By the early Bronze Age, in the fourth millennium BCE, a few of these cultures had developed writing, allowing accounts to be kept and

The entrance of the Great Temple at Abu Simbel in Egypt. (Wikimedia Commons/Than217)

complex ideas to be recorded in a durable way. In Mesopotamia, in China, in India, in Egypt, empires emerged. From these empires came the earliest calendars.

Calendars most likely became necessary for many reasons in an increasingly complex society. Priests needed to know in advance when to perform religious ceremonies to appease the gods. Taxes and tithes had to be calculated correctly and collected promptly, especially from the remoter parts of the empire.

Nature provided three cycles upon which the first calendars were built. The day and the year were the basic units of time. Between them, the phases of the Moon allowed the year to be divided into smaller, more convenient periods, and most early calendars were based on lunar months, even though the Moon itself has no direct biological impact on humans.

However, the solar year could not be divided into an exact number of lunar months. Twelve lunar months are about eleven days shorter than a solar year. Early astronomers devised many ingenious solutions to this puzzle, most of them involving the insertion of an additional month every few years at a time determined either by observation or by the application of formal rules. The simplest of these rules added an extra month every three years, or three months every eight years. By the fifth century BCE, Babylonian and Greek astronomers had discovered the Metonic cycle, a remarkably close (but entirely

At the Chinese New Year, it is customary to give gifts of money in small red envelopes. (Wikimedia Commons/BCody80)

accidental) equivalence between 19 solar years and 235 lunar months. In the Babylonian calendar, an extra month was inserted in years 3, 6, 8, 11, 14, 17 and 19 of each cycle. The Metonic cycle remains important today in the Jewish calendar and in the formula for calculating the date of Easter.

The civil calendar of Pharaonic Egypt was unusual in the ancient world. From very early times, it ignored the Moon completely and had twelve months of thirty days each, followed by five days to bring the length of the year to 365 days. This was a closer approximation to the solar year than 12 lunar months, but the seasons still slipped by a day every four years in this calendar, making one complete cycle in 1,461 years. The Egyptians did not try to fix this slippage by adding leap days. Instead, they used the first appearance of Sirius in the morning sky in early summer – called its helical rising – to anticipate the start of the annual flooding of the Nile, an event of immense importance to their civilisation.

The Gregorian calendar is the civil calendar used in most of the world today. It has its origins in ancient Rome of the first century BCE, as the Republic gave way to imperial rule. Rome's earliest calendar was lunar, but for at least the four later centuries of the Republic, the calendar was based on ten months whose lengths were alternately 31 and 29 days, with an eleventh month (*Ianuarius*) of 29 days and a twelfth month (*Februarius*) of 23, 24 or 28 days. There was also an intercalary month called *Mercedonius*, of either 27 or 28 days, which was appended after *Februarius* in eleven years of a 24-year cycle. This complicated scheme gave an average year length of 365.25 days, which was quite close to the true length of the year as defined by the seasons.

Nevertheless, by the middle of the first century BCE, the entire calendar had slipped by three months relative to the seasons. When Julius Caesar became *Pontifex Maximus*, he used his authority to order a reform of the calendar. Three intercalary months were inserted to bring the calendar back into its traditional correspondence with the seasons, the lengths of the twelve regular months were adjusted so that they added up to 365 days rather than 355, and a day was added to every fourth year to maintain an average year length of 365.25 days. The emperor Augustus later made further corrections because the priests had applied the leap year rule incorrectly, but from 8 BCE onwards, the Julian calendar was used unchanged for almost sixteen centuries, surviving the fall of the Roman Empire and becoming the calendar of Christianity.

The most important event in the Christian year is Easter, which commemorates the death and resurrection of Jesus. It seems to jump around the calendar at random, falling anywhere between 22 March and 25 April in an apparently unpredictable way. There is, however, a reason for this. Easter is

Brightly painted Easter eggs are a tradition in many European countries. These are from Romania. (Pixabay/jackmac34)

tied to the phases of the Moon, making it the one lunar event in an otherwise solar calendar. The drama of the first Easter, as told in the Gospels, took place over the Jewish festival of Passover, which begins on the night of the first Full Moon of Spring. Over several centuries, the direct link to Passover was broken, but a formula was eventually agreed for the date of Easter: it is the Sunday after the first Full Moon which falls on or after the Spring equinox, which is assumed to be 21 March.

The mathematics required to accurately calculate the date of the Full Moon would not be available to astronomers until Isaac Newton developed his theory of gravity and invented the calculus, in the late-seventeenth century. Most medieval tables of dates of Easter therefore used the 19-year Metonic cycle to set the date of the Full Moon, and this cycle is still the basis of the formula for calculating the date of Easter today.

By the sixteenth century, it had become apparent that the calendar of Julius Caesar contained a serious flaw. The average length of the year, 365.25 days, was slightly longer than the true length of the year defined by the passing of the seasons, 365.2422 days. The difference amounted to less than 11 minutes each year, but as the centuries passed, astronomers noticed that the Spring equinox was slipping slowly backwards through March. The phases of the real

Moon had also drifted away from those predicted using the Metonic cycle. The result was that Easter was being celebrated on the wrong day.

The matter was addressed at the Council of Trent, sitting between 1545 and 1563, which authorised Pope Paul III to reform the calendar. Expert advice was sought from astronomers and mathematicians. Proposals included an ingenious method for adjusting the Metonic cycle to keep it in step with the real Moon over many centuries. This was developed and championed by the Jesuit Christopher Clavius, who was a leading mathematician at the *Collegio Romano*.

The reform was promulgated in February 1582 by Pope Gregory XIII in a papal bull titled *Inter gravissimas*. It was in three parts. The first was the omission of ten days in October 1582, to bring the Spring equinox back to 21 March. The second was a modification of the leap year rule. Century years that were not divisible by 400 would no longer be leap years. This removed three leap years every four hundred years, and made the average length of the year equal to 365.2425 days, just sixteen seconds longer than the true length of the cycle of the seasons. By this rule, the year 1600 would still be a leap year, so nobody living at the time of the reform would notice any difference from the old calendar.

The third part of the reform was a subtly modified Metonic cycle. Corrections were applied in some century years to stop the phases of the calendrical Moon drifting away from those of the real Moon. The resulting formula for calculating the date of Easter is both elegant and accurate, and it is simple enough that a ten-year-old with a calculator can use it.

Most of the Catholic countries of Europe adopted the new calendar in 1582 or very soon afterwards, but in Protestant England, the reform was viewed with deep suspicion at a time of heightened religious sensitivity. Britain did not adopt the Gregorian calendar until September 1752, by which time it had become necessary to omit eleven days to bring the equinox back to 21 March. The English tax year had traditionally started on 25 March, the old New Year's Day, so that date was also shifted eleven days in 1753, to 5 April. In 1800, when a leap day was omitted from the calendar under the Gregorian rule, the start of the tax year was moved again, to 6 April.

The French Revolution of 1789 toppled the Bourbon monarchy which had ruled the country for five centuries. Most histories of the Revolution focus on the execution of Louis XVI and Queen Marie-Antoinette, and on the five years of political turmoil which culminated in the Reign of Terror. It is less well known that the Revolution also gave birth to a new calendar, known as the French Republican calendar. This was intended to supersede the Gregorian calendar, which was seen as part of the hated *ancien régime*.

The new calendar replaced the seven-day week with a ten-day period called a *décade*, and each month contained three *décades*. They were given names reflecting Nature and the seasons. The first month of winter, for example, was called *Nivôse*, meaning "snowy", whilst summer began with *Messidor*, derived from the Latin *messis*, meaning 'harvest'. Years were denoted by Roman numerals. Thus the French almanac office, the *Bureau des Longitudes*, dated its founding to a law passed by the *Convention National* on 7 *Messidor* an III, which corresponds to 25 June 1795 in the Gregorian calendar.

Five or six days were added at the end of each year to bring its length to 365 or 366 days. This practice was very similar to that of ancient Egypt. However, in keeping with the belief that the Revolution should replace religious superstition and tradition with scientific rationality, the start of each New Year in the Republican calendar was defined to be the day on which the true autumnal equinox occurred at the Paris Observatory. This could not be determined using a simple leap year rule. It required the true ecliptic longitude of the Sun to be calculated to determine the exact moment when its value was 180°. In an age before computers, this was a laborious task, and it was one of the reasons why the Republican calendar was abandoned after just thirteen years. Another was that ordinary people felt duped at only getting one rest day every ten days rather than every seven. And, of course, the rest of the world declined to adopt the new calendar, in spite of the scientific superiority asserted by its creators.

Every year, the lives of almost two billion people change drastically for an entire month. They abstain from all food and drink between sunrise and sunset. This is Ramadan, the month of fasting, one of the five Pillars of Islam. Its start and end are determined by sightings of the crescent of the New Moon. The Islamic calendar is purely lunar, so each year sees Ramadan begin ten or eleven days earlier relative to the seasons. When Ramadan falls in the northern hemisphere's mid-summer, as it did in the 2010s, the daily period of fasting can last for sixteen hours or more for Muslims living in the higher latitudes of

An artistic depiction of the month of Floréal (April/May) in the French Republican calendar. (Wikimedia Commons/Louis Lafitte (artist), Salvatore Tresca (engraver))

Pilgrims circle the Kaaba at Mecca during the annual Hajj. (Wikimedia Commons/Zakaryaamr)

Europe and North America. By 2030 it will begin in mid-winter, and the daily fast may be as short as eight hours.

The Moon also triggers the largest annual mass movement of humans on the planet: in late January or early February, hundreds of millions of Chinese people travel from the great industrial cities to the towns and villages where they grew up, in order to celebrate the Chinese New Year with their families. The New Year begins on the day of the New Moon which falls between 21 January and 20 February in the Gregorian calendar.

In India, and in the Indian diaspora across the world, the first New Moon of autumn, in October or early November, marks Diwali, a celebration of the triumph of light over darkness which is observed by Hindus, Jains, Sikhs and some Buddhists.

Thus, the cycle of phases of the Moon touches the lives of more than half of humanity every year, something that might easily be overlooked in a world where most human activity is tied to the daily rising and setting of the Sun and the turning of the seasons.

The Serenity – an ancient, multi-species alliance that rules the cosmos for the benefit of all. Here the Supreme Council, the most powerful beings in the entire galaxy, meets to discuss an unusual survey report – a small planet with some very strange creatures living there. (John McCue)

Report on Planet Three

Philip Wallace

Deep in the galactic core floated a mammoth structure: a space station, nay, an artificial world, a megastructure on a vast scale. This was the headquarters of the Serenity, the multi-species alliance that ruled the entire galaxy. Born in war and conquest in ages past they now followed more peaceful and enlightened principles. For five thousand generations they had kept peace and prosperity – for the most part – across the entire galactic domain.

In one particular chamber was gathered a collection of twelve beings of varied shape and size. This was the Supreme Council, the body that set policy and made all the vital decisions. Today they met, assembled in a half-circle as usual, to discuss the latest reports from the Surveyors, who scouted out new worlds and civilisations to assess whether they were suitable for contact.

The leader of the Council, a being called Magladoroth, leaned back in his chair, his six eyes glancing up at the vast curved window that formed the far wall and the magnificent sight beyond it, the supermassive black hole around which all things in the galaxy turned. These survey meetings were always tiring, and often contentious, but today was an exception.

"I must congratulate you Chief Surveyor," he spoke, his voice low, gravelly and rumbling like a distant avalanche. "I believe that is the smoothest running survey meeting we have ever had. One hundred and sixty-eight new worlds and all of them approved for contact to become valued additions to the Serenity. Well done."

The Chief Surveyor looked pleased at the praise, but also slightly shifty. "I beg your pardon Your Supremacy but there is actually one additional planet. One I am…conflicted about presenting to this Council."

The six-eyed being focused on the Chief Surveyor while his eleven fellows groaned, sighed, or otherwise expressed their frustration that the tedious meeting wasn't over yet.

"And why might that be Chief Surveyor? What can this new planet be like that makes you concerned about our reactions?"

The Chief Surveyor squirmed under the many-eyed gaze. Of course since its species was gelatinous "squirming" was a completely normal action, but this squirming was something more.

Our survey team thought it a beautiful world – from orbit at least. Closer analysis was … concerning. (NASA/Apollo 17/ AS17-148-22727)

It sighed. "The planet occupies the third orbital position around an otherwise typical yellow K'Zatz-type star. It is on the outer edge of the Life Zone of the star and possesses one major natural satellite about one-quarter its mass. The mineral composition is again standard, a fairly standard iron/nickel molten interior, and a rocky surface mostly comprised of various metals and silicates. About 70 percent of the surface is covered in liquid water, with large ice caps at the poles. Surface terrain runs from deserts to tundra, coastal plains and mountain ranges. Fairly typical."

A brief pause as the Chief Surveyor gathered its thoughts. "Plant and animal life is abundant in all varieties expected on such worlds. Analysis of accessed records indicate some growing concerns over a loss of biodiversity and a potential imminent mass extinction but there is a lack of consensus among the dominant species. As for them…Perhaps, Your Supremacy, it would be best if the Council all reviewed the survey report. I am confident that you will understand my doubts. I will then answer any questions you have. I should clarify that the cultural information and the supporting imagery is largely obtained from accessing their global data network – which is for the most part almost pathetically protected against outside intrusion."

The six-eyed being bobbed his head in agreement and the Council members began to read, watch, listen, smell or telepathically absorb the report as their species preferred. Magladoroth idly reached up with two of its three-fingered claws to scratch at the curved horns that flanked its head, a subtle indication of mild irritation among its species. It did not take long before the first startled question came from the large spider-shaped Councillor, the voice hissing.

"They really called their planet "Dirt?""

The Chief Surveyor grimaced; an impressive action considering it lacked a face. "That is how the native name translates into Serenity-Basic, Councillor."

That answer produced a few mutterings before the Council continued reviewing the report. Ten minutes passed in near-silence, the only sounds being a quiet rumble of the space station's systems and the occasional susurration of breathing – from those Councillors that actually breathed, that is.

Magladoroth set down the report. "I see why you had concerns Chief Surveyor. The planet itself appears to be standard enough as you said, reasonably fertile….but there are an exorbitant number of contaminants present. Can you account for this?"

The yellow-green gelatinous Chief Surveyor, Hoother, shrugged again. "I believe that is explainable by the dominant species present on the world. Humans, they call themselves. They have only recently undergone industrialisation and entered the computer/information era, a process which involved, and still involves, burning a vast amount of fossil fuels to achieve."

The giant blue spider, known to the others as Webspinner as no-one else could pronounce his true name, hissed once more.

"They actually burned them? Those fools, did they not know the dangers to their biosphere from such unbridled consumption?"

Hoother turned to face the large spider. "This is what is so concerning Councillor Webspinner. Apparently they *do* know the dangers, or at least have known of them for the last fifty years or so, but they continue to use them anyway."

Magladoroth shivered. "Are we sure these are the dominant species? It is difficult to believe a group so ignorant could dominate a world."

Hoother bobbed slightly in acknowledgment. "The humans are certainly the ones building and maintaining all the structures. However an analysis of their global data networks and the more common elements thereof suggest that the *actual* dominant creatures might be a pair of four-legged species apparently called "cats" and "dogs." Many humans are utterly devoted to caring for these creatures and there is often vehement disagreement over which species is superior."

Councillor Starlight, a gaseous entity that looked like a small floating nebula, flared briefly, small white points twinkling within the blue-white core and the red outer layer, its voice echoing in the minds of those listening.

"Are these….traces of radioactive isotopes in their atmosphere? Did they have some sort of atomic war?"

Although appearing to be in near-constant conflict, examples of what may be the actual dominant species on this world, known as "cats" and "dogs". (Wikimedia Commons/Peretz Partensky/ flickr.com/people/68877611@N00/Attribution-ShareAlike 2.0 Generic (CC BY-SA 2.0))

Hoother bobbed again. "They have developed atomic weapons, yes. But to their credit they only used them in warfare twice."

Starlight did not sound impressed. "From the trace levels of radiation they must have built very powerful weapons for only two of them to have been used."

Hoother could only grimace again. "Only two were used in *combat*, Councillor. Many hundreds were detonated after that as "tests" – a euphemism for demonstrations to intimidate their rivals."

The twelve Councillors all recoiled in horror at the idea. Councillor Starlight in particular was aghast. "They *deliberately* detonated atomic weapons *in their only biosphere* to intimidate others? What kind of monsters are we dealing with here?"

Hoother shuddered slightly, indicating agreement for the most part. "Again, to their credit Councillor, many humans opposed such policies and advocated strenuously for complete atomic disarmament. They have made considerable progress in that regard – current global inventories of atomic weapons are a fraction of what they were only a few decades ago."

Starlight flared briefly again – this time in the yellowish region between the core and outer layer, signifying considerable distress – before continuing. "If they have atomic weapons then it logically follows they must have atomic power. Are they at least using that on a large scale in an effort to reduce or eliminate fossil fuel consumption?"

Hoother shook itself. "No Councillors. Atomic power is used on a limited scale, more so in some nations than others, but there is considerable public

"The humans detonated high-yield nuclear weapons as tests. This is most concerning." (National Oceanic and Atmosphere Administration/United States Department of Energy)

opposition to it. Apparently it is not considered "safe" – despite vast amounts of scientific data showing that it is."

Webspinner hissed again. "Large-scale atomic power is a necessary post-industrialisation measure before practical fusion and vacuum power generation can be developed. Do these humans *want* to progress beyond their polluted world?"

Hoother turned to face the spider. "Many of them do want to progress beyond their world Councillor, and colonise another, nearby planet. Many others oppose them, if only because the main proponents are exceedingly wealthy, rather arrogant and somewhat difficult to deal with, hence they are often disliked on principle."

At that Councillor Yun'Gak'Ret reared up to it's full one and a half metre height, its two red compound eyes focusing entirely on Hoother. The six green limbs trembled slightly in agitation as the giant insect began clicking its mandibles together. The various systems in the Council chamber automatically translated for the benefit of non-insectoid beings.

"The more I read of this report the less I want to know. Section Sixteen–Alpha, Collective Psychology, is particularly concerning!"

The other Councillors quickly reviewed the section in question, before incredulous laughter (or the species-specific equivalent) echoed through the room.

Webspinner gave voice to the joke, its bright scarlet rear section vibrating in an approximation of laughter. "How can such a large proportion of these humans think their planet is *flat*? And another significant fraction thinks their early forays into interplanetary flight were a hoax. Even worse, another significant fraction, seemingly entirely separate from the rest of the prior categories think something called "Astrology" can be used to predict their personal futures! And this most recent nonsense – vaccines are a way of tracking people and communications emitters give you an infectious virus? By the stars, what is *wrong* with these beings?"

Another round of laughter followed that outburst, though it was quieter and more contemplative than before. Councillor Starlight then asked Hoother how this "astrology" was meant to work.

Hoother shuddered. "Allegedly, Councillor, the relative positions of their primary star and more distant constellations allows people's futures to be mapped out, with suggestions and warnings against certain actions. In practice, it relies on basic psychological tricks, highly generalised statements and educated guesswork. I do not have any data on how many humans *genuinely* believe in this method, but information about it, and "horoscopes", are freely available."

Webspinner continued. "I would oppose any kind of Contact with this species on that basis alone, never mind the rest of the appalling mess they seem to make of things."

Hoother spoke up. "Respectfully Councillor, whilst those views and these "Conspiracy Theories" are indeed ludicrous, they are very much the minority views. Most of the population, in these respects at least, is more intelligent."

Starlight brusquely cut in. "They may be smart enough not to believe such nonsense, but they are still lacking. They not only burn fossil fuels for power, and somehow consider at least some of it to be "green" but their most common forms of transportation, be it ground vehicles, skycraft or waterborne vessels all burn even more fossil fuels. They actually drive around in ground vehicles powered by continuous small explosions. Nuclear pulse propulsion may be a viable idea for spacecraft but not personal vehicles in an atmosphere."

Magladoroth once again re-entered the conversation, its lower-right arm holding the report while the upper pair of arms once again scratched its horns in irritation. "Section Twelve, the Biosphere Impact Analysis…this shows the planet is rapidly heading towards an ecological catastrophe, with increased global temperatures, melting ice caps and extreme weather events becoming commonplace. Surely the humans know of this pending cataclysm?"

Hoother grimaced yet again (it had grimaced more in this meeting than the previous ten meetings combined, it mentally noted).

"Their scientific community is well aware of the pending crisis and has in fact been attempting to warn their political leaders and the general public about it for several decades, with limited apparent effect. The public are largely apathetic, or have been until relatively recently, while the politicians are not interested in anything beyond their next election."

"Devastating wildfires are becoming more common as the biosphere deteriorates." (Kaibab National Forest, Arizona/Attribution (CC BY 2.0))

Yun'Gak'Ret clicked angrily again: *"You said the public were apathetic until recently. What has changed?"*

Hoother turned to face the insectoid. "More and more people are becoming aware of the dangers and are engaging in "activism" and "raising awareness" of the issue. They are also some groups engaging in protests of…questionable effectiveness. Battery-powered personal vehicles are also becoming more popular, as are hydrogen-fuelled models."

Webspinner asked the obvious questions. "Activism? Raising awareness? What does that even mean Chief Surveyor? And what of these "protests?""

Hoother checked its notes. "A large portion of the population that is active on the global data network is apparently under the impression that simply talking about an issue, and perhaps buying a battery-powered vehicle is all they need to do. This does, of course, overlook where the electrical energy for that battery comes from in the first place – or the materials for that matter."

Another glance at the notes followed. "As for the protests, there was one group that attempted to protest fossil fuel use by…attaching themselves to road surfaces with adhesive. Another group tried another protest by, um, burning their fossil-fuel engines and moving very slowly. We do not have full details on the intention behind that one as yet. A notable feature of these protests and "activism" in general is that it is the younger generation that is most involved."

Starlight pulsed briefly. "So the younger generation may be able to stabilise the situation?"

Hoother grimaced yet again. "Possibly, Councillor. However the younger generation also have a habit of paying vast sums of money to travel long distances around the world on something called "holidays," using the same fossil-fuel burning skycraft and waterborne vessels they protest against. Even "celebrities" – humans with some degree of notoriety as entertainers – who have spoken out on environmental issues do so, and are often used in advertising to promote those very same long-distance journeys. I have no explanation for this almost schizophrenic situation."

The Council was once again silent. Magladoroth sighed. This was not looking promising for humanity.

"What of their conflicts? Section Five mentions they are not even close to a united people."

Surprisingly it was Councillor Dives-Into-Sunset, a large golden birdlike being that hung upside-down from the ceiling that answered. "I have considered this aspect at length while other matters were discussed. In many respects this is a primitive and violent culture. Wars for resources have happened frequently, as one might expect from recently-industrialised worlds. But they also fight for conquest, for ideology, for religious doctrines and even for the *wounded pride* of their leaders. This is a species that will happily kill another group simply for following the wrong religious text even if it's notionally *the same God*."

"Skycraft travel is extensively used for "holidays" – even by those who protest climate damage." (Wikimedia Commons/Patrick Hawks)

Webspinner shuddered, an impressive sight considering its six-metre leg-span and four-metre height. "This is not a civilised world. In one of their larger countries they had a brutal civil war because one sub-faction wished to keep another group of humans as slaves. While most humans now agree slavery is abhorrent – as they should – it is deeply concerning that this is such a comparatively recent development, and that such bloodshed was needed to end it."

Hoother could only bob in agreement. "I cannot dispute those statements Councillors. This species has fought many long and vicious wars for occasionally futile, idiotic or asinine reasons."

Magladoroth shuddered again, all four of its claws clenched. "So they fight each other for stupid reasons, they continue to ruin their biosphere despite knowing what they're doing, then protest it ineffectually. The few trying to spread out into their star system are either mocked or under funded in the case of the scientifically-based efforts. They are not a united government; many states have only a passing approximation of democratic rule…and in the recent past have enslaved each other in mass numbers."

Hoother grimaced yet again. "An apt summation, Your Supremacy."

Starlight flared again, regret lacing its mental tone. "Perhaps we should refer this world to the Planetary Protection Division to be quarantined and potentially sterilised."

"Warfare has been endemic on this world, causing tens of millions of deaths in the last century alone." (Wikimedia Commons/Kenneth Forbes/Canadian War Museum/19710261-0142)

Everyone shuddered at the blunt suggestion. Quarantine of a newly-discovered world was relatively common, but it was mostly reserved for worlds where the dominant species was still developing from the hunter-gatherer stage. Certainly it was almost unprecedented to suggest it for a world this populous and this (apparently) advanced.

Sterilisation was even worse. It had not been done since the wars that founded the Serenity five thousand generations ago, and that was done to a species that was a clear threat to all life in the galaxy.

Webspinner reared up at the suggestion. "You cannot be serious Councillor Starlight. They are primitive, stupid and violent, I agree, but they cannot *possibly* be a threat like the Sar'Gat were."

Starlight replied in an even tone, not even slightly fazed at Webspinner's vehemence. "They are not such a threat *now* Councillor, but if they encounter other species they could learn knowledge and technology that would be dangerous to all. I would support Quarantine at the least."

At this point, a new voice entered the conversation. Councillor Haranghast raised a hand to be recognised. One of Magladoroth's eyes saw this and idly waved a double-jointed arm in his direction, giving permission.

"Councillor Starlight raises a valid point, one that we may already be too late to prevent." The voice was as high-pitched as one might expect from such a small creature – for Haranghast was barely a metre tall.

Magladoroth focused all six eyes on the diminutive being. "Please elaborate for us Councillor."

"Of course, Your Supremacy. According to the data gathered, there are many reports of extra-planetary visitors, known as "UFOs," particularly in the last century. Many, indeed most of these accounts, are easily explained as hoaxes, hallucinations or liars seeking attention – but there are enough common elements to concern me. Are we *absolutely sure* there has been no prior contact?

Councillor Webspinner answered that. "The last survey ship to enter this system did so a thousand years ago. No survey or patrol vessel has ever approached closer than the planet's moon, as per standard protocols for uncontacted species."

Haranghast nodded. "Then it may simply be a coincidence. After all, the common descriptions of UFOs as "flying saucers" or "cigars" are not even close to a fair description of our vessels, and the descriptions given of the "extra-terrestrials" don't match any Serenity species. There are a number of folklore elements though that suggests some knowledge. My own species appears in their myths; they call us "centaurs" though they think we're much larger." There was a rippled of amusement at that. "There are also myths with a passing resemblance to the species of Councillors Yun'Gak'Ret and Webspinner, but

"The most common human depiction of extra-terrestrial life. Obviously mistaken, none of us look like that!" (Wikimedia Commons/Luke Hancock/Creative Commons CC0 1.0 Universal Public Domain Dedication)

given that smaller, non-sentient versions of those species appear endemic on this world that may simply be human imagination running wild."

Haranghast took a breath and smoothed down the brown fur on the lower portion of his body. "Councillor Starlight's concerns remain valid however. I too would support a Quarantine."

Hoother was horrified at how quickly the meeting had turned to such grim topics. Magladoroth noticed the gelatinous blob's anxiety.

"Chief Surveyor, do these humans have *any* redeeming features?"

Hoother settled slightly in relief. "I believe so Your Supremacy. There is a significant portion of the population that recognises the issues and is actively working to improve them. Their scientific community has worked wonders; virtually eradicating several diseases that were previously endemic. Childhood mortality rates have dropped massively in just the last century. Major efforts are underway to improve the ecological situation. They are early efforts in many cases, and are hampered by political apathy and public misunderstanding, but they *are* making progress."

Another glance at the notes followed in a brief pause. "Additionally, in cultural areas they produce some genuinely impressive works of art, ones that even the Galactic Gallery would be happy to have on display. Their body of literature is wide, varied and similarly contains some truly impressive works, tales of epic deeds and personal triumphs. They have, uniquely for a single-species world, created a vast array of musical genres – enough that every species present here would find something pleasing."

The last image taken of Planet Three by the withdrawing survey ship. According to the human data network this is almost identical to a famous image inspired by one of their more open-minded and forward-thinking scientists, a "Carl Sagan." We can only hope more of the population listen to those like him – before drastic action is required." (NASA/JPL-Caltech)

Hoother broke off then, with an approximation of an embarrassed smile. "And having tested various samples that were brought back, their collection of alcoholic beverages is…staggering in a literal and figurative sense."

Magladoroth looked amused at that. "Chief Surveyor, did you really just present *varieties of alcohol* as a redeeming feature of a primitive culture?"

"You asked for redeeming features, Your Supremacy, and that is one of them."

There was another round of amused laughter before Magladoroth called the vote. Given the tone of the prior discussion it was hardly a surprise when Magladoroth presented the Council's final verdict:

"This species is dangerous and primitive. Contact is forbidden. They are to be discreetly observed and we will re-assess them in one hundred years…if they are not extinct by then of course."

It's Not in the Stars

Why Astrology has no Place in the Modern World

Neil Haggath

"Astrology does prove one thing – that there's one born every minute!"
Sir Patrick Moore

Suppose someone was to tell you that they believe the Earth to be at the centre of the Universe, and that the Sun, Moon and planets revolve around it, all attached to crystal spheres and turned by angels. You would surely think they were a lunatic! But the difference between believing that and believing in astrology is … what, exactly? It can be argued that there isn't any, as astrology originates from a time when people really did believe in an Earth-centred Universe!

Astrology, in various forms, has been part of human culture for thousands of years, and for much of that time it was taken seriously, even by the intelligent. Even kings and political leaders used to consult astrologers about affairs of state and military campaigns. Until the sixteenth century, astrology and astronomy were thought to be related, and many eminent pre-telescopic astronomers, such as Tycho Brahe (1546–1601), believed in

Tycho Brahe, the last great pre-telescopic astronomer, was a meticulous observer and catalogued the stars with unprecedented accuracy. Yet he believed in an Earth-centred universe and in astrology. (Wikimedia Commons/ Eduard Ender)

astrology. But it's all a hangover from an ancient geocentric world view which has been obsolete since Galileo first turned a telescope to the sky.

Yet an alarming proportion of people today still profess to believe in it to varying degrees – from those who just casually read 'their stars' in a newspaper to those who actually allow it to influence decisions in their lives, and even in business and politics. Some believe that people of certain 'star signs' are 'compatible' with each other, while others are not, and choose whom to date and marry on the basis of their 'sign'. Even more outrageously, it is not unheard of for employers to select candidates for jobs on the same basis!

Some people try to defend astrology by saying something along the lines of, "It has been around for thousands of years, so there must be something in it!" That is a ridiculous non-argument; *why* does it follow that just because it's very old, it must be right? On the contrary, we know it to be absurd, precisely because it's very old! Astrology originates from a time when people lacked the benefits of modern knowledge, and believed that the Earth was the only 'world', that the Sun, Moon and planets were 'lights in the sky' which revolved around it, and that everything had been created by God or gods purely for the benefit of Man.

But that world view ceased to have any validity in the seventeenth century. First, Galileo's telescopic observations showed that the Moon is a world, with mountains, plains and craters. Astronomers then proved that the Earth is *not* the centre of the Universe, but is one of the planets which orbit the Sun. And finally, Isaac Newton and his contemporaries formulated the modern laws of physics, and showed that the motions of celestial bodies could be explained by physics and mathematics.

The famous Dendera Zodiac, a sculpture from the ceiling of a chapel in the Egyptian Temple of Dendera and believed to date from around 50 BCE, depicts the signs of the Zodiac and other Egyptian constellations. It is now displayed in the Louvre in Paris. (Wikimedia Commons/ Gary Todd/CC0 1.0 Universal (CC0 1.0))

The kind of astrology with which we are familiar in the western world originated in the Mediterranean region, and particularly Egypt, around 2,500 years ago – ironically, about the same time as the classical Greeks discovered the principles of real science! It is based on the concept of 'signs' or 'houses' of the Zodiac, which are equal 30° sectors of the Ecliptic – the Sun's apparent path around the sky in the course of a year, with respect to stars and constellations. (The ancients, of course, thought the Sun really *did* travel around the sky, and did not know that it was actually the Earth that moved). The signs have the same names as twelve already existing constellations – some of which are believed to date from thousands of years earlier – with which they roughly coincided at that time – 'roughly', as the constellations are *not* of equal sizes. But they no longer do, due to the phenomenon of precession.

The direction of the Earth's axis of rotation slowly changes direction, like that of a tilted gyroscope, with the North Celestial Pole tracing a circle on the

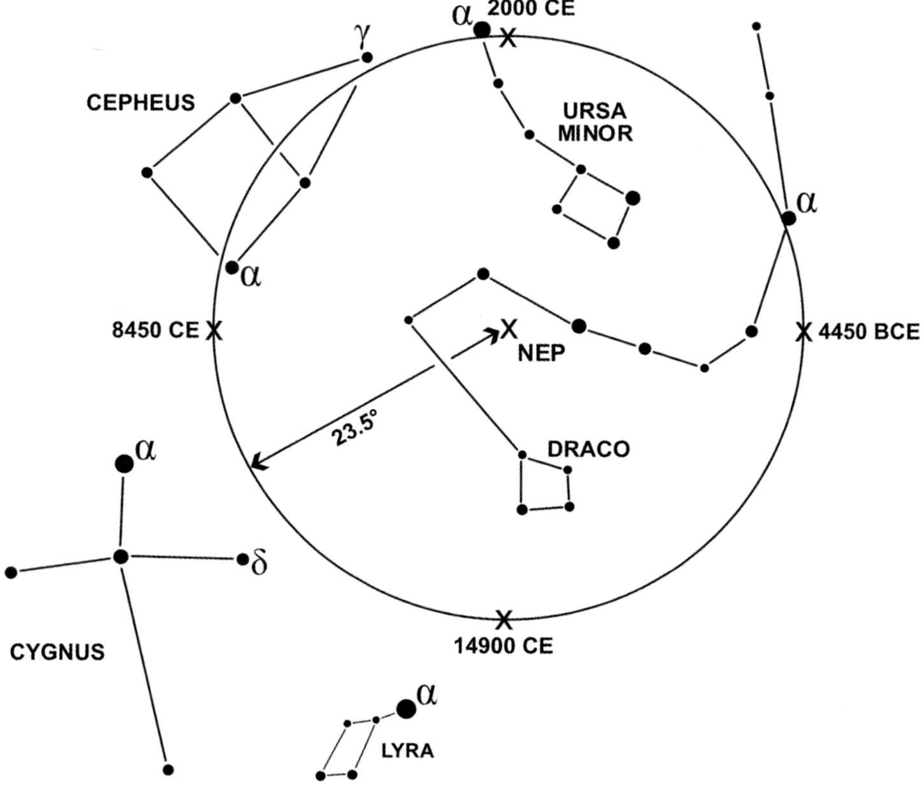

The precession of the Earth's axis causes the North Celestial Pole to follow a circular path around the sky, known as the Precessional Circle, in a period of around 25,800 years. The position of the Pole is shown for various epochs past, present and future. (Neil Haggath/Garfield Blackmore)

sky in a period of 25,800 years. This also means the Celestial Equator changes position with respect to the Ecliptic, and the Vernal Equinox, the point from which the 'signs' are defined, moves around the Ecliptic in that same period. Consequently, the 'signs' are now 'out of step' with their eponymous constellations by roughly one; when astrologers say the Sun is in the 'sign' of Gemini, it's actually in the constellation of Taurus, and so on, and the Vernal Equinox, also known as the First Point of Aries, is in the constellation of Pisces. (In many ancient cultures, the Vernal Equinox was regarded as New Year; this is why Aries is regarded as the first 'sign' of the Zodiac.) In fact, they move out of step by one 'sign' roughly every (25,800/12) or 2,150 years.

For this same reason, we know that astrology could not have originated earlier than about 1000 BCE; had it done, then Taurus would be the first 'sign', and the 'signs' would now be out of step by *two* constellations.

The simplest form of astrology, that which forms the basis of the daily 'horoscope' columns in newspapers, assumes that a person's 'star sign' – the 'sign' which the Sun was in on the date of their birth – is all that matters. Your 'sign' somehow determines your personality, and what will happen in your life on any given day. This seemingly makes the utterly ludicrous assumption that all human beings can be divided into twelve groups, and that everyone in each of those groups has the same personality, and their lives all follow the same course! On a given day, every person in the world whose sign is, say, Taurus, will be 'lucky in love', or will have to 'worry about money matters', or some such.

Naturally, the 'predictions' in these columns are always masterpieces of vagueness, worded in such a way that anyone can read something into them, if they want to. Two things which *are* 'predictable' are that they don't really predict anything, and that those in different papers, on the same date and for the same 'sign', are not likely to say the same.

Furthermore, as most people will only bother to read it for their own 'sign', the authors can write variations of the same thing for several 'signs', and get away with it. I once looked at one such column to check this out, and found that it included, for *every* 'sign', some variation on the theme of, "If you're single, you will find new love." So *every* person in the UK who was single would supposedly 'find new love' on the same day!

Naturally, all this stuff is designed to make the 'believers' feel good about themselves; nobody wants to be told that they are going to have bad luck, do they? Indeed, from 1994 to 2000, the weekly televised draw for the UK National Lottery was preceded by a (some would say) utterly nonsensical astrological 'prediction' about what kind of person might win.

The supposedly 'serious' practitioners of astrology will agree that the newspaper variety is over-simplified. They claim that the exact date and time of a person's birth is important, rather than just their 'star sign'; some claim that their place of birth is also important. They perform personal 'readings' for people – for a fee, naturally – much as so-called 'clairvoyants' do, and use charts, called horoscopes, which 'map' the positions of celestial bodies to imaginary 'areas of influence', or some such nonsense. (This is the correct meaning of 'horoscope', rather than the, "If you're a Taurus, this will happen to you today" in newspapers – not that it makes any more sense!)

I have no idea what these people actually say or write, but I would bet any money that they employ the same sort of trickery as so-called 'clairvoyants' or 'psychics', who claim to 'foresee the future', or apparently deduce things about their clients, which they could not possibly know. All such apparent abilities depend very much on 'cold reading' – manipulating a conversation so as to lead the subject to give away personal details, without realising that they are doing so. Once again, their 'predictions' of the future are phrased in such deliberately vague terms that they can mean just about anything; you can *find* a meaning, if you *want* to believe it!

The sixteenth-century French astrologer Michel de Nostredame (1503–1566), better known by his Latinised name of Nostradamus, wrote a book of 1,000 poetic verses (all but 58 of which survive), *Les Prophéties* (The Prophecies), which claimed to predict future events. Modern day 'believers' claim to have interpreted his work as having 'foreseen' all manner of events in modern history, including the rise of Napoleon and Hitler, both world wars, the nuclear bombing of Hiroshima and Nagasaki and 9/11. In fact, his 'prophecies' consist of vague and meaningless gibberish, which can be interpreted or misinterpreted to mean almost anything one chooses. Strangely – or not – modern authors nearly always claim that he 'predicted' whatever important event has just happened at the time of their writing!

Astrologers of this kind also claim to produce 'personal profiles' for their clients, which are supposedly complete descriptions of people's personalities, based on their time and place of birth. I have no idea how these are supposed

The sixteenth-century French astrologer Nostradamus, seen here in a portrait painted by his son César de Notre-Dame, is said to have predicted … just about anything you want to believe he did! (Wikimedia Commons/César de Notre-Dame/Sasha I)

to work – but not surprisingly, they are always written in the same sort of vague terms, such that their subjects can interpret them however they choose.

Most astrologers make no attempt to explain *how* the positions of the Sun, Moon and planets in the sky can possibly influence people's lives or personalities; it obviously has nothing to do with the laws of nature or the principle of cause and effect, and perhaps everything to do with mysticism and something akin to magic. A few, however, have made pathetic attempts to justify it in vaguely 'scientific' terms, which only serve to prove their utter ignorance of science!

A well-known British astrologer once stated on TV: "The Moon exerts a gravitational pull on the water in the oceans, to cause the tides – so it *must* have some effect on the fluids in the body!" Naturally, he made no attempt to say *how* or *why* any 'gravitational pull' on the fluids in the body should influence a person's personality – or why, given that that pull is always present and ever-changing at every moment of a person's life, its particular configuration at the moment of birth should be so important!

I can only assume that this person never learned even the most elementary physics at school, and has never even heard of the inverse square law!

Firstly, it is a complete fallacy to say that one object 'exerts a gravitational pull on' another. Every two objects exert a *mutual* gravitational pull *on each other*, which depends upon *both* of their masses! The law of universal gravitation, as formulated by Isaac Newton, states as follows: "Every object attracts every other object, with a force which is directly proportional to the product of their masses, and inversely proportional to the square of their distance apart."

So the *mutual* gravitational attraction between the Moon and the water in the Earth's oceans is proportional to the masses of both. The mass of fluid in a human body is very much less than the mass of water in the oceans – by a very rough estimate, it's smaller by a factor of about 100 million million million – so the attraction between *it* and the Moon is correspondingly smaller. In other words, it's absolutely miniscule. Note also, "inversely proportional to *the square of their distance apart*" – if the distance is doubled, the force is reduced by a factor of four, and so on. The Sun, Moon and planets are very big masses, but they are also a very long way away.

In fact, it's very easy to prove that at the moment of a baby's birth, the gravitational influence of the doctor or midwife is comparable to that of Venus (the closest planet) or Jupiter (the biggest), and greater than that of any of the other planets. It's much smaller than that of the Moon – but astrologers claim that the influence of the planets is equally as important.

Of course, it isn't necessary to know *how* something is supposed to work, in order to prove that it doesn't! Numerous statistical studies have proved that the accuracy of astrological 'predictions' is no better than what would be expected due to pure chance. In one such, a 'respected' astrologer was asked to write horoscopes for a number of people, while a psychologist also prepared profiles for them. A number of other astrologers were sent three psychological profiles and one horoscope, and asked to match the horoscope to one of the profiles. Their success rate was 34% – almost exactly what would be expected from random guessing.

Many other studies have shown that those who *want* to 'believe' will do so, no matter what; they will believe that a horoscope or profile is their own, if someone tells them it is! The American illusionist James Randi – famous for replicating the feats of so-called 'psychics' and 'clairvoyants', while never claiming to be anything but an illusionist – once visited a college, posing as an astrologer, and pretended to write personal horoscopes for each student in a class. He then asked them all to rate the accuracy of 'their' horoscopes, and the majority claimed them to be accurate. Randi then told them to pass the horoscopes around, revealing that they were in fact all identical. The single horoscope had been written in the usual vague terms, such that those who wanted to 'believe' could read into it whatever they liked.

Australian researcher Geoffrey Dean conducted a test using 'real' astrological readings, which were done by an astrologer for a number of 'believers'. Half of the subjects – the control group – were given their horoscopes as they were; for the other half, Dean 'reversed' the horoscopes, substituting phrases which were the opposite of what the original said. Of those in the second group, just as many – 95% – claimed that the readings applied to them as those who had been given the 'correct' readings!

But my favourite example is that of French psychologist Michel Gauquelin in the 1980s. He got an astrologer to write the horoscope of a certain individual, and then sent identical copies to 150 'believers', telling each one that it had been written for them, and again asked them to rate its accuracy. 94% of the subjects said they recognised themselves in the

The Canadian-American author and scientific sceptic James Randi (1928–2020), was a renowned exposer of charlatans, including astrologers. (Wikimedia Commons/James Randi Educational Foundation/Eduardo Aparicio/GNU FDL/ CC BY-SA 4.0 DEED)

description. The actual subject of the single horoscope was in fact … an infamous serial killer![1]

Many people will no doubt think, "OK, so we know astrology is a load of rubbish, but what harm does it do?" At first sight, you may well think that if some people want to believe in a ridiculous unfounded superstition, then it's up to them, and it doesn't do any harm to anyone else. But I beg to differ.

Many years ago, I was horrified to find that W. H. Smith, one of the UK's biggest chains of bookshops, had books on astronomy and astrology randomly mixed on the same shelf – not by accident, but as deliberate policy! It's easy to imagine the potential harm, if well-meaning but naive parents had bought their children the wrong book. Thankfully, my complaint to their head office led to them rectifying the matter.

As I mentioned earlier, it's not unheard of, even today, for employers to choose between candidates for jobs on the basis of their 'star sign'! Does anyone think *that* is harmless? Think about that; in most of the civilised world, it's now illegal to deny anyone a job on the basis of their gender, race, religion or sexual orientation – but you can still get away with doing so on the basis of an irrational archaic superstition!

Even worse, there have been cases of people in real positions of power being influenced by it. In 1988, it was revealed that the dates and times of meetings in the White House were often arranged and rearranged on the advice of an astrologer! It was actually Nancy Reagan who consulted the astrologer, but her husband – the man in the most powerful political position in the world – went along with it, and allowed matters of state to be influenced by random gibberish.

Adolf Hitler is known to have consulted an astrologer, before making political or military decisions. At this point, I expect some apologist to point out that Winston Churchill also did so during the War, and *he* was no lunatic! Indeed he did – but not because he believed any of it; it was because he knew that Hitler did, and wanted an insight into the sort of nonsense which was being fed to the latter. (I suspect, however, that the Führer's astrologer simply told him whatever he wanted to hear!)

As recently as 2015, a British Member of Parliament stated his belief that astrology can somehow be used to diagnose illnesses – I don't pretend to understand how! – and actually proposed that it could be used to take pressure off doctors in the National Health Service. This is a man who is paid to play a part in the running of the country, and who might actually have a say in such things as the funding of the NHS!

---

1. 'Your Astrology Defence Kit', Andrew Fraknoi, *Sky and Telescope*, August 1989.

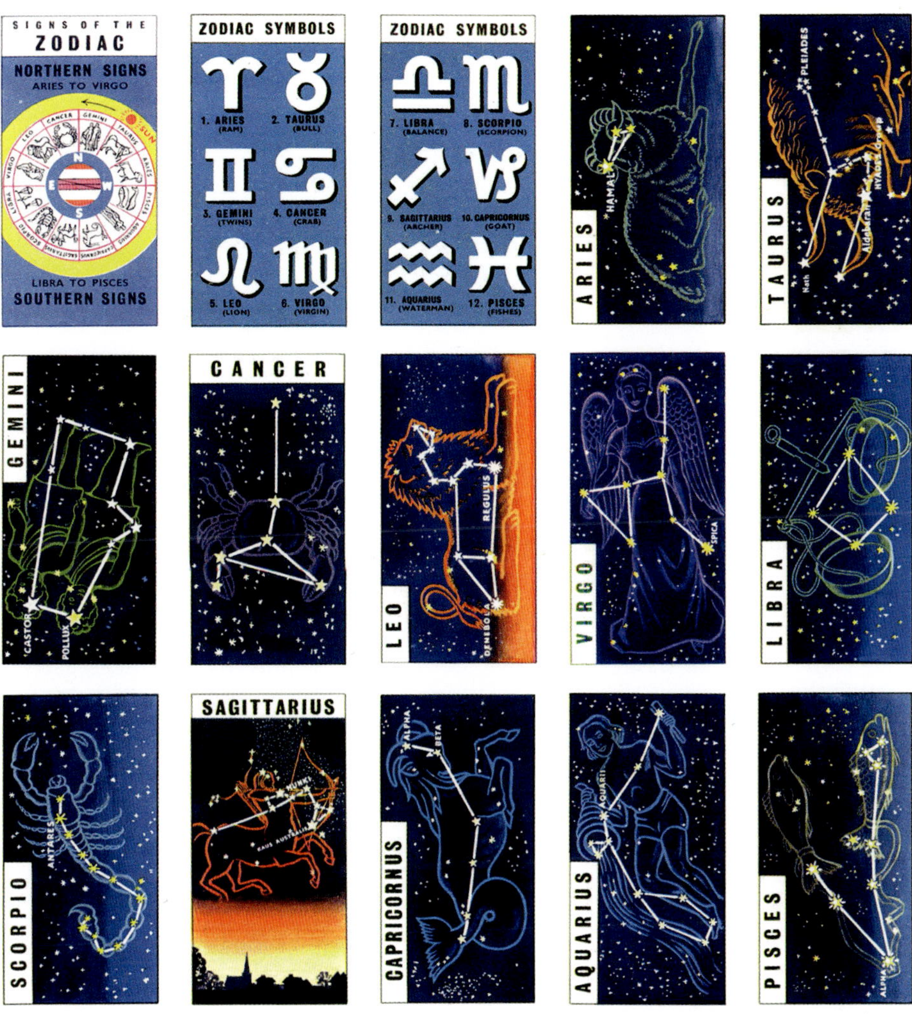

In March 1956 (and reprinted in 1958), Brooke Bond & Co. Ltd. issued a set of 50 collectible cards on the theme of astronomy, titled *Out Into Space*. Given away in packs of tea, most of the cards in the set depicted the Sun, Moon, planets and constellations, together with astronomical phenomena such as solar and lunar eclipses. Strangely, however, the set mixed astronomy and astrology; the 15 cards shown here depict, as well as the 12 constellations of the Zodiac, a horoscope chart of the astrological 'signs', and their symbols. (Lipton Teas and Infusions/ Brooke Bond & Company/ekaterra)

But on top of all this, astrology is just one of many examples of what has come to be called 'anti-reality'. It promotes irrational, anti-scientific thinking, of a kind which has no place in the modern world.

Other kinds of anti-reality are becoming ever more widespread in today's society, and poisoning people's minds – creationism, global warming denial, anti-vaccination madness (especially so in the time of Covid-19), and no end of barmy stories about imaginary 'government conspiracies'. In today's world, science is vital to every aspect of our daily lives – even for those who are too naive to realise it, while happily enjoying its benefits in the form of their computers, mobile phones, satellite TVs and so on – yet people's minds are being turned against it.

Astrology may be relatively harmless, in comparison to some of those other things, but it's still a part of the general trend of anti-reality. After all, those who believe in one kind of delusion are probably likely to be swayed by others. It plays a part in weakening people's ability to use logic and think rationally, at a time when this is more important than ever.

Finally, as an astronomer, it irritates me that astrology blinds people to the *real* wonders of the Universe. The Cosmos is fascinating and wonderful in its own right, without having to make up fairy tales about it! In an age when men have flown to the Moon, we have robot rovers exploring the surface of Mars, and we have sent space probes to the most distant planets and beyond, people still cling to ancient fantasies, left over from the time when our ancestors thought those bodies revolved on crystal spheres.

We teach children to believe in magic and fairy tales – but then they grow up. It's time that we all, as a species, did exactly that!

Astronomy and Science Fiction

David A. Hardy

Since the dawn of humankind, the stars and planets were just mysterious lights in the sky, some fixed, some moving. The Sun and Moon were gods, to be worshipped. Then, in 1608, the Dutch spectacle-maker Hans Lippershey turned the first telescope on the skies. A year later, the Italian astronomer Galileo Galilei did the same, with better results, and it was realised that there are other worlds out there – perhaps with life on them? By 1865, the French novelist Jules Verne had written *From the Earth to the Moon*, a fanciful story in which his heroes were fired into space from a giant cannon, while in 1900, the English writer H. G. Wells used anti-gravity, all thanks to a material called 'Cavorite'. The idea of travel to other worlds had well and truly entered the public mind. The concept was to be eventually realised in 1969 with the Apollo 11 Moon landing – and the future is now wide open!

*　*　*　*　*

The nineteenth century French novelist, poet, and playwright Jules Verne, whose adventure novels included *From the Earth to the Moon*. (John McCue)

Originally published in 1865 by French editor and publisher Pierre-Jules Hetzel (1814–1886) *From the Earth to the Moon was* later translated from the French by Louis Mercier and Eleanor E. King and released by Scribner, Armstrong & Company, New York in 1874. The cover of this edition is seen here. Jules Verne's epic tale was not the first work of fiction to describe a journey to another world, although it was probably the most widely-read. (Wikimedia Commons/Jules Verne/ PD-US/Smithsonian Libraries/SIL-SIL28-090-01/ Scribner, Armstrong and Co., New York)

When did humankind become aware that we live on a planet, and that our planet was just one of many? It was during the second century CE that Claudius Ptolemaeus of Alexandria (also known as 'Ptolemy') produced the first map of the world. Although distorted and inaccurate by modern day standards, many places depicted on the map, such as Great Britain, can be easily recognised. Ptolemy knew quite well that the Earth is a sphere, and he also knew a lot about the five bright 'wandering stars' or planets. His only real (albeit natural) mistake was in assuming that Earth is the centre of the universe. So mankind had made much progress since King Menes was on the throne.

Polish astronomer and mathematician Nicolaus Copernicus (1473–1543), seen here in a portrait from Town Hall in Toruń and dating from c1480 by an unknown artist. (Wikimedia Commons/ Toruń Regional Museum/PD-Art/ Unknown Artist)

Nicolaus Copernicus proved that the Sun, and not the Earth as originally thought, was at the centre of the solar system. The steady emergence of Copernicus' heliocentric theory – in which the Earth and planets revolve around the Sun – into society eventually brought about a significant shift in the way that humanity viewed the universe. (© David A. Hardy/astroart.org)

Yet the Egyptians themselves had contributed little to the growing river of knowledge. They had observed the heavens, but had made little progress in astronomy; they believed that the Earth was flat, and to them the stars were just objects to be used as points of reference. The fact that the Egyptian priests paid great attention to the positions of the stars is shown by the orientation of the great pyramids that still stand in the desert. These seem to have been used partly as observatories, but also as temples and partly as tombs. The most famous of them is the Great Pyramid of Cheops, built around 3000 BCE. The date of its construction can be estimated with some certainty, because its main passage points almost directly towards what was then the north celestial pole. The shift of this polar point is very slow. It describes a circle in the sky, with a period of around 26,000 years.

Over the centuries the power and glory of Egypt had been replaced by the more brilliant glory of Greece. Other civilisations that had grown, then decayed, were the Babylonians, who thought that the Earth was a floating disc; the Indians believed that the entire universe was enclosed in a hollow iron sphere; while the Chinese studied eclipses and left records of 'dragons and hairy stars'. So, two main ideas retarded the scientific progress of the ancient world: that the Earth was flat, and that it was the centre of the universe. It had taken two thousand years before the work of Copernicus proved convincingly that the Sun, not Earth, is the centre of our System.

The idea that the Earth is flat is natural enough. Above us is the vast sky, with its blazing Sun by day and its twinkling stars by night. (But why does the Moon change its shape night by night?) Below, the solid ground stretches in all directions, and extends far beneath our deepest wells and shafts; while to all sides the horizon meets the sky. There is no obvious curvature. Later Egyptians believed the Earth to be rectangular, with Egypt in the middle, deserts and seas all around. Xenophanes, a Greek philosopher born in around 570 BCE, taught that the Earth is a true disc, with its upper side touching the air and its lower side extending without limit.

Many scientists were unhappy with the theories of the day. We don't know for sure who took the final step, but we do know that Aristotle wrote, '… the sphere is the shape a body naturally assumes when all parts of it tend towards the centre.' This suggests that he may have had the first glimmerings of an understanding of gravity; though two thousand years were to pass before Isaac Newton discovered the true laws of universal attraction.

Aristotle also made the point that the altitudes of stars change according to the location of the observer. Polaris is higher up in Greece than in Egypt, because Greece is further north, whereas the bright southern star Canopus never rises at all in latitudes north of the Mediterranean. Polaris itself remains invisible from latitudes south of the equator. This cannot possibly be explained if the Earth is flat, but is only to be expected if it is spherical.

Then there is the Moon. It had been realised for many years that it has no light of its own, but depends entirely upon reflected sunlight, like a great celestial mirror. So when it passes into the shadow of Earth, all direct light is cut off from it, and Earth's shadow can be seen upon the Moon's disc – a lunar eclipse. It was Aristotle who pointed out that since this shadow is curved, Earth's surface must also be curved.

It was fortunate for science that Aristotle took this view, because his opinions were accepted almost without reservation; right up to the time when the Tudors ruled England. If Aristotle said the Earth is round, then round it must be. The only remaining problem, then, was to determine its size. It took another Greek, the astronomer and mathematician Eratosthenes of Cyrene, to do this, during the third century BCE. Eratosthenes' method of finding the size of the Earth was simple, but completely sound. It was based on the fact that when the Sun is directly overhead in one place, it is some way from the overhead point in another. Eratosthenes chose Alexandria, on the shore of the Mediterranean, where he lived, and Syene (the modern Aswan), some distance to the south on the banks of the Nile.

So far we have talked about stars. If you watch these for a number of nights you will see that each star stays in its own place. After a number of nights

Speculation is rife as to what planets outside our solar system, or exoplanets, look like. This image reveals a variety of possibilities. Since the discovery of the very first exoplanet was confirmed in 1992, a total of over 7,000 additional exoplanets have come to light. This increasing abundance of worlds beyond our own continues to fuel the idea that humanity is not alone in the universe and, when this is eventually proved to be the case, human culture will undergo what is undoubtedly the biggest shift it could ever experience. (NASA/JPL-Caltech)

you may notice that a particular bright star, which may shine steadily instead of twinkling like the rest, seems to have got closer to or further away from another star that was near it. This is a planet. The name comes from the Greek word *planetes*, meaning 'wanderer'.

Through a telescope, a star will still look like a point of light, although a planet will show a definite disc. The Earth is just one of nine major planets orbiting the Sun. And the Sun is just a star, which just happens to be much closer than the other stars, which is why it looks round and is so bright. We now know that there are many, many more 'solar systems' out there, as we have discovered 'exoplanets'. These can take many forms, and there must surely be many worlds that are similar to Earth, and which contain life, which could take many forms, including similar to humans. Sadly, we may never know, as it takes just over four years for light to reach us from even the nearest star, Proxima Centauri (a member of the Alpha Centauri system). Stars are usually many thousands of kilometres across; some vastly bigger than our Sun, some much smaller. The hottest ones shine with a bluish light, while our Sun is yellowish and the cooler stars are red.

A drawing, made in black India Ink and coloured inks, intended to bring together many of the (perceived) elements of spaceflight at that time; Earth, Mars, Venus, the Moon, a ferry rocket, spacecraft and astronaut. (© David A. Hardy/astroart.org)

The Earth travels around the Sun at about 107,000 kilometres per hour – much faster even than, say, an Apollo rocket! If it could suddenly stop moving it would fall into the Sun, because this, being 1.39 million kilometres across, has a much stronger gravitational pull. It was only natural that once we found out about the true nature of our universe, people would want to try to find ways to leave our planet and visit others. The Moon, being greatly closer, was the obvious first target. But it also quickly became obvious that (despite trying!) hot air or gas-filled balloons, and then even aircraft were not the answer.

As mentioned previously, giant cannons seemed a possibility, and in 1865 the French novelist Jules Verne (1828–1905) wrote *From the Earth to the Moon*, in which a hollow projectile was fired from a 900-foot-long cannon at such a speed that it would escape Earth's gravity. (Actually the crew would be squashed flat in the first second). A sequel to this story was *Around the Moon*; the explorers never landed, and Verne believed (correctly) that the Moon was lifeless. In this he was unlike H. G. Wells who, in 1901, wrote *The First Men in the Moon*. His spaceship is coated with 'Cavorite' (named after the eccentric scientist, the fictional Dr. Cavor), which repels gravity, and his insect-like Selenites live in great underground caverns. These were of course some of the earliest examples of what we now call science fiction.

The Moon captured the imagination as the first possible destination should humankind find a way to break Earth's gravitational field and achieve spaceflight. (© David A. Hardy/astroart.org)

So any vehicle to go into space needs to leave our atmosphere, and then somehow overcome gravity. Balloons and aircraft helped scientists to learn a lot about the atmosphere, although this is a small problem when compared with the mysterious force of gravity. This is the force which tries to pull everything to the centre of Earth (indeed, any large object in space attracts a smaller one in the same way). For anything to leave Earth forever it was found that it would need to achieve a speed of 40,000 kilometres per hour. A giant cannon is clearly of no use, and there is as yet no such thing as antigravity.

From about 1880, the Russian schoolmaster-turned-rocket-scientist Konstantin Tsiolkovsky (1857–1935) worked on flying machines. Later he

Russian schoolteacher turned scientist – and the founder of theoretical astronautics – Konstantin Tsiolkovsky. (Wikimedia Commons)

tried to design spacecraft, and realised that only rockets could power these. He also decided that solid fuels were not powerful enough; so he invented a rocket using liquid fuel, pumped from tanks. Even more important (as we now know well) he suggested putting one rocket on top of another, so that when the bottom one was empty it would fall away, reducing the weight, while the next one took over, and so on. All of this is now familiar to us: we call it a *step-rocket*.

The Austro-Hungarian-born German physicist and engineer Hermann Oberth (1894–1989) did not know about Tsiolkovsky's work. In around 1924 he also designed rockets to fly into space, and even wrote a book about his work entitled *Die Rakete zu den Planetenräume* (The Rocket into Planetary Space). This book sold well, and created a lot more interest in space travel. Oberth also worked on an early German film about a trip to the Moon, and tried to build a rocket for it – which exploded. Societies began to be formed, especially in Germany, and amongst the many other space travel related popular books that were written was *The Conquest of Space* by the German-American science writer Willy Ley (1906–1969).

Meanwhile, the American engineer and physicist Robert Hutchings Goddard (1882–1945) had written a book about studying the upper air by rockets, in which he added that a rocket could be sent to the Moon carrying

On 16 July 1969, the Apollo 11 astronauts were launched from Kennedy Space Centre atop a Saturn V rocket. Propelled into space at seven times the speed of sound, the Apollo missions carried with them the dreams of humanity. (NASA)

flash powder so that we could see its arrival. This caused several newspaper reports, and again led to a lot of people take an interest in the subject. Goddard went on to build liquid-fuelled rockets, which by 1935 were attaining speeds of over 1,100 kilometres per hour and rising to heights in excess of 300 metres.

A few years later there was another war. Goddard's ideas had not been taken very seriously in America, but German scientists, including Wernher

Human culture has truly embraced the fact that science fiction often becomes science fact. Aside our own Earth which, by comparison, is dwarfed by giant planets such as Jupiter and Saturn, there are great numbers of much smaller worlds known as asteroids, or minor planets. Science fiction has often proposed their mining in the future, with science fact determining that the likes of giant metal-rich asteroid 16 Psyche is awaiting such excavation. (© David A. Hardy/astroart.org)

von Braun (1912–1977), had continued with their experiments. When the war broke out they were ordered to build rocket-bombs to be fired at England. In 1944 the first V2 rockets began to fall around London. After the war several of the German scientists went to America, along with some captured V2s. There, in a large desert area called White Sands, New Mexico, the explosive warheads were replaced with scientific instruments. In December 1946 a V2 rose to just over 180 kilometres. Soon the Americans had no more V2s, so they built their own rocket, called Viking. They also put a small rocket, the WAC Corporal, on top of a V2, making the first step-rocket, which attained a height of around 400 kilometres. By 1955 the Americans announced that within two years they would put into orbit an artificial satellite – *Project Vanguard*. Thus began the chain of events which resulted in twelve men walking on the Moon, manned spacecraft going into orbit, and many unmanned spacecraft visiting other

Science fiction frequently guides human imagination in the direction of actual scientific research and development. Examples include close-up views of other planets which, although currently available only from visiting space probes, often enter the science fiction genre, as illustrated here. Onboard the *Liberator* space ship, the rebel leader Roj Blake (played by Welsh actor Gareth Thomas), who headed up a mutinous crew for BBC TV's science fiction series Blake's 7, views a Jupiter-like planet from the bridge. (© David A. Hardy/astroart.org)

planets, even going into orbit around them. Science fiction has not only kept pace with each new development, but has often exceeded them; the concept of visits to other worlds, even different dimensions and universes, has well and truly entered popular culture, and indeed is often regarded as standard fare.

How Do Birds Find Their Way?

John McCue

How does a young barn swallow, born in the United Kingdom in early summer, find its way without adult help to Africa, thousands of kilometres away? There are still mysteries to uncover about the fascinating skills that migrating birds possess, but the answer to that first question has partly been found in displacement experiments. Perhaps unjustly, we can trap young birds as they set out on their first journey, move them to another location and then release them – they are unable to correct their flight and end up somewhere else! We deduce that birds must be born with an innate knowledge of their species' direction and duration of

Barn swallow. (John McCue)

travel – they know which way and for how long – but have never been that way before. Birds actually have better eyesight than humans during the day and on their first flight undoubtedly learn to follow certain geographical marks, such as rivers and coastlines. Honey buzzards, for example, fly from West Africa to Europe using the familiar shape of Italy as their Mediterranean crossing point, and many other species have well-known and well-worn routes. The author has friends with racing pigeons and these birds are the best-known example of landscape followers. Tracking these pigeons as they repeat races shows that they gradually perfect their route home, and so post the quickest time, the aim of the sport! It's heart-warming, nonetheless, to know that naïve young birds are not usually kidnapped and taken elsewhere. Adult birds will often lead them – cranes and geese are regularly seen flying in formation, like the Red Arrows – so that the juveniles pick up the routes from their elders. Other major routes for migration are Turkey, where billions of birds cross each year, and Cyprus, where, sadly, many passerines – small birds – are trapped to feed the human culture of restaurant meals. Songbirds are a special delicacy. Thankfully, the considerate side of human nature, aware of the global life cycles of our planet's creatures, is coming to the rescue and reducing this needless slaughter.

Apart from watching the ground, are there other methods that birds can use?

We all know what time of day it is, but the reader may not be aware that birds do also! Without this built-in clock, birds would find it impossible to traverse their long migratory journeys in search of food and breeding grounds.

Honey buzzard. (John McCue)

The longest of these routes are undertaken by seabirds, led by the impressive Arctic Tern. During northern summer, this tern breeds in the Arctic regions of northern Europe, heading south for Antarctic regions during northern winter, a return journey of about 30,000 kilometres. Amazingly, as they fly south, then back north again, they see almost constant daylight, more than any other creature on the planet.[1] This illumination by the sun also helps them to see fish just below the ocean's surface. The Common Tern is more leisurely, usually breeding over wider areas of Europe and wintering in South Africa, but one unfortunate chick that was ringed in Sweden carried on, we suppose, from South Africa and was found freshly dead in New Zealand, five months after its departure[1]. If it had lived and returned to Europe, it would have covered an exhausting distance of almost 50,000 kilometres, easily more than the girth of our planet earth. The sooty shearwater, another seabird, holds the top spot though, for the longest regular migration. These record-breakers breed in the New Zealand summer before heading north up to Alaska, exploring the Pacific Ocean for food in a huge figure-of-eight, finally clocking up around 65,000 kilometres over the return trip, at an average of over 300 kilometres a day! They also breed on South Atlantic islands near Antarctica, and then likewise set off northwards, feeding and exploring the Atlantic Ocean, arriving

---

1. RSPB, *Bird Migration*, Ian Newton, William Collins, 2010, p163.

around Britain's coast from July to September. Their sightings from Britain are scarce, indeed they are classified as 'near threatened', but at least they are regular visitors.[2]

Sailors have long known the importance of time in helping them find where they are. Pinning down your latitude is easy; just look at the Pole star, Polaris, in the constellation of Ursa Minor (the Little Bear), with a sextant. Read off the star's altitude above the horizon and that's your latitude, immediately. The sun's altitude at noon, its highest point during the day, is also helpful with

Sooty shearwater. (John McCue)

latitude. However, that's only half of the story; a ship also needs its longitude – without that it could be anywhere along the latitude line, and heading for a loss at sea or a crushing shipwreck on unexpected rocks. To be sure, before the early eighteenth century, such maritime havoc was common. The final straw came on the night of 22 October 1707. The fleet of Admiral, Sir Cloudesly Shovell failed to spot the Scilly Isles and was destroyed with the loss of 2,000 men. The British government was finally persuaded, in 1714, to offer a cash prize, equivalent to one-and-a-half million pounds in today's value, for a solution to finding the longitude of a ship at sea. Birds have a similar issue, though not with such disastrous consequences! Finding their north-south position, i.e., latitude, is straightforward but not so their east-west longitude.

So how does time help? If we in Britain want to call a friend in Australia, we well know that their time is ahead of our own, so we can allow for this difference and call at a congenial time. Likewise, we must be lenient in phoning the United States, which is in the western, and opposite, direction. Their time is behind UK's. This simple idea was pursued tirelessly by a working Yorkshire joiner called John Harrison. His life was the antithesis of those of the lauded academic astronomers in London, who were convinced that the answer lay in the heavens. Harrison's dream was to build a clock that would carry London's time with the vessel. The navigators would then wait for noon, when the sun is at its highest point. Looking at the London clock would then offer up the difference in time between the ship and London, hence how far east or west the ship was

---

2. *Britain's Birds*, Rob Hume et.al., Princeton University Press, 2016.

of the UK capital. Problematically, timepieces were not very advanced in those days, the most accurate being the pendulum clock. Such a clock at sea was as useful as a billiard table!

Undeterred by such thoughts, Harrison's chronometers, slowly but surely, became smaller and more accurate as his labours progressed. After forty diligent years, he produced his master timepiece which nestles in the palm of the hand, and he was awarded the prize in 1773. Three years later, in perfect synchronicity, he died on his 83rd birthday.

Birds do not, we need hardly state, know the time of day as accurately as Harrison's chronometers, but they are innately aware of the day's passage. On a sunny day, Common Starlings

John Harrison, horologist who made the first chronometers that were accurate enough to measure longitude at sea. (John McCue)

kept in a large circular cage pointed their beaks in the same compass direction, despite the sun moving across the sky on its daily journey. In practice, if an Arctic Tern for example decides, one morning, to migrate directly south, and it can see the sun, it knows it must aim to the right of the sun. The tern, being aware of the progression of the day, will know to fly *towards* the sun around noon time, and then later in the day, the bird must make haste with the sun on its right. What is more impressive is that birds can detect the polarisation of sunlight as well, without wearing Polaroid sunglasses! It's long been known that light waves consist of two parts, an electric field and a magnetic field, vibrating at right-angles to each other. Normally, sunlight is unpolarised, that is, the two field vibrations are orientated in random directions (though still fixed at right-angles to each other), like wave ripples on strings all angled in different directions. A wooden (polarising) fence would then prevent nearly all the strings' ripples passing through to the other side, only letting through the waves matching the fence slots. Now sunlight scattering from the molecules of the air is polarised, and particularly strongly in the direction at ninety degrees to the original travel of the sunlight – in other directions of scattering the polarisation is only partial. Consequently, if we look up in the sky at sunset (or sunrise), we will see strong polarisation (if we wore Polaroids!). Birds can detect this polarisation and use it as a signpost.

As mentioned, birds cannot easily find longitude, and if they are subject to strong winds, drastic action is needed. The author was lucky enough to see an eastern crowned warbler in Trow Quarry, South Shields, on the northeast coast of England, in October, 2009, apparently blown off course from its usual stomping grounds in China! It was spotted by two local birdwatchers who gained

Eastern crowned warbler. (John McCue)

the credit for the first recorded sighting of this species on British land. Another marathon east-west trip was reported by the Royal Society for the Protection of Birds: an American Wigeon was found lost and wandering in the UK, but managed to find its way back across the North Atlantic.[3] The only recourse such unfortunate birds have on long east-west journeys is their sense of day-time – this tells them if the sun is rising a little bit earlier than they expected. If such be the case, they know they must be flying eastwards, and vice-versa if they see a later sunrise than they anticipated. The clear benefits flowing from north-south migration – plentiful food and good breeding grounds – are easy to understand in terms of different seasons in the northern and southern hemispheres, and the author speculates that maybe birds are thankful for this since east-west travel is relatively more difficult!

But what of the night? How does the bird then proceed in the right direction? With *our* astronomical knowledge, we know that as our planet sweeps around the sun we see different star vistas from the night side throughout the twelve months – we see Orion in the northern winter and Cygnus in the northern summer. Birds that migrate during the night are able to learn which stars are visible at their regular month of departure. This skill builds on the birds' sense, not only of the passage of the day, but also of the months of the year. Experiments[4] show that birds kept in an internal environment for months, during which time their days and nights provided by artificial lighting, can still keep track of their yearly natural cycle. They know when it's migration time (they put on weight), breeding time (they sing a different song), and moulting time (they obviously shed feathers). This endogenous yearly body clock helps them remember the stars they see on their seasonal migratory way. It makes sense that the birds' main star method is the circumpolar motion of the stars

3. Search for 'Dangers of Migration' on the RSPB website **rspb.org.uk**
4. RSPB, *Bird Migration*, Ian Newton, William Collins, 2010, p259.

during the night. Those stars are there all the time, of course, not rising or setting as they rotate around Polaris in the northern hemisphere and indeed that stationary pole star is the perfect guide to flying north or south. These ideas about birds' star knowledge are confirmed in a planetarium. If the night sky display is actually moved round by, say, 180 degrees, the birds' flight will change by that amount also. Moreover, if the display is configured to make Betelgeuse the pole star, the birds will follow that star instead. In the southern hemisphere though, there is no bright star near the south celestial pole, so it is worth speculating that the Sooty Shearwaters in New Zealand know the patterns of stars, the constellations indeed, in that part of the sky to show them the northerly direction.

What if the sun and the stars are not options for the birds? The earth's magnetic field will certainly suffice: its lines conventionally flow from south to north, and so if the direction of those lines can be detected, the south-north route is evident. Interestingly, the magnetic field is generated, in a reverse dynamo way, by large, swirling electric currents in the earth's outer liquid metallic core. The effect is that of a monster bar magnet inside the planet – hence the field lines must ultimately drop vertically into the earth's surface near the north pole cap (and rise vertically from the southern cap). The major issue then, for polar birds like the Arctic tern, is that field direction, south-north, becomes much more difficult to ascertain. To be fair, this downside is balanced by the advantage of the gradual downward dip of the lines as they pass beyond the equator (where they are horizontal) on their way to the poles; the angle of dip being greater further from the equator. If the bird can detect this angle of dip, it will tell them their latitude, roughly.

There are two ways that birds may detect the magnetic field: firstly by light-sensitive photo-pigments in the bird's retina (though, strangely, they are believed to exist only in the bird's right eye). They contain cryptochrome, a protein, which reacts to a blue sunlight photon by transferring an electron between its two constituent molecules. It's the only protein known to do this. The molecules then flip between parallel and anti-parallel spin, the amount of flipping giving the bird information (transmitted to its brain by the optic nerve) about the magnetic field. This is believed to give directional information to the

Arctic tern. (John McCue)

bird, the compass. Secondly, magnetite, a mineral with iron content, has been found in the base of a bird's beak, and this substance responds to the dip of the surrounding magnetic field, giving latitude to the bird. Crucially, the magnetite is connected by a cranial nerve to the bird's brain. Fish, amphibians, and other creatures also have these magnetoreceptors.

Although, of course, amphibians such as frogs, toads, and newts have evolved to survive on land they still must breed in water, but at least they don't have as far to travel as birds. For example, most frogs and toads only crawl a few hundred metres to their breeding pond or lake, so detecting the earth's magnetic field is not necessary – so some amphibians can do this, some cannot. Those species that do not have magnetoreceptors use mainly olfactory and visual clues to find their route. Salmon migration is the most impressive of fish journeys. The Atlantic salmon feeds on herring but must breed in more suitable waters – fresh, highly oxygenated, and shallow. They swim thousands of miles, some to Britain, to find these waters; they strike up rivers, clearing any obstacles to find a place to breed. Some fish have been seen leaping three metres up a waterfall. The young mature after two or three years then return to the ocean. Crucially, they then find their way back to the same breeding stream because they know for certain that it's a good place and they do it firstly by using magnetoreceptors. Once they have found their home river, they use the memory smell from their youth to rediscover their stream. It is an extremely arduous process, and some salmon die from exhaustion once they have reached base.

The popular North Atlantic minke whale, seen in British coastal waters, migrates towards the tropics in the winter for breeding, and heads for its summer feeding grounds in the Antarctic. Many whale species undertake similar journeys, but do they have magnetoreceptors like the salmon? Jesse Granger

Minke whale. (John McCue)

of Duke University, North Carolina, may well have solved this puzzle.[5] She studied sunspot numbers and found that they matched up with the numbers of whales found beached. She only counted those whales that had stranded themselves while fit and healthy, to eliminate those who had died while ill or injured. Sunspots are patches where the sun's magnetic field lines become tangled up like the line on a fisherman's reel when it runs away with itself. This wound-up area prevents energy from directly below reaching the surface, causing a drop in temperature of 1,000 degrees. This is not much compared with the Sun's overall surface temperature of 6,000 degrees, but it is enough to make the sunspot look dark by comparison – if seen in isolation, a sunspot would be bright. All this is the fault of the sun's varying rotation at different latitudes. But how could these sunspots possibly push the whales off course? Sunspots are symptomatic of highly energetic surrounding areas where flares could occur. The sun's surface is usually a boiling cauldron of bubbling energy anyway, but a flare is a huge explosion that sends a raging stream of energy and particles across the solar system, sometimes impacting the earth. The energy covers the entire wavelength range of the electromagnetic spectrum, from radio waves to X-rays, and Granger suggests that it is the former which confuses the whales, with tragic consequences. The shorter of these radio waves, often called microwaves, are used across all communication systems, from TV masts to satellite earth-links. On further scrutiny, she found that correlation with strandings was even clearer when these radio waves from the sun were at their strongest. Maybe then whales *do* have magnetoreceptors, and they are disrupted by this radio noise, throwing the creatures off their intended journeys, like swiveled road signs. The sun's blast of charged particles also affects the earth's magnetic field lines, unsurprising since the connection between electricity and magnetism is well known even across all our daily lives; and some researchers think that not only are the poor whales' magnetoreceptors twisted out of line by the solar storm, but the earth's magnetic field is slightly deviated by the charged particles. No wonder the whales suffer mayhem!

Bird migration is the example noted in all our minds of impressive journeys in the natural world, but butterflies come a close second, particularly the painted lady. No other British butterfly has this rosy-orange upper side and black wing-tips marked with orange,[6] but this beautiful creature cannot survive the British winter. Where they go, or even if they survive at all, was not known

---

5. *The Atlantic*, Ed Yong, 2020, **theatlantic.com/science/archive/2020/02/link-between-solar-storms-and-whale-strandings/606910**
6. *Collins Complete Guide to British Insects*, Michael Chinery, Collins, 2009.

until 2009, when citizen science stepped in. Ten thousand butterfly lovers took part in that year's survey of this well-loved insect by Butterfly Conservation,[7] and with the help of many more European sightings they found that painted ladies arrive from Africa in spring, but radar sightings showed that they leave us for the return journey at a high altitude of about 500 metres, beyond the sight of

Painted lady. (John McCue)

observers – they can reach speeds of up to 50 kilometres per hour at this height. No wonder we didn't know to where they disappear in the autumn. At the end of each breeding summer, of course, many more leave than arrive. Many studies of the monarch butterfly, an equally impressive flyer across North America, and counterpart to the painted lady, have shown that, just like birds, they can use the sun to guide them, given their innate sense of the passage of the day. In cloudy conditions, the monarch has a back-up – it follows the direction of the earth's magnetic field, detected by cryptochromes, presumably in the butterfly's head. Birds also use this method as mentioned earlier. The monarch studies reach a greater level of detail however, by pinpointing the precise range of wavelengths that activate the cryptochrome; 380 billionths of a metre up to 420 billionths, which is in the blue-to-ultraviolet region. American butterfly aficionados will no doubt be generous in allowing that the painted lady may well beat the return journey of the monarch (around 6,500 kilometres) by a factor of two. Some of the painted ladies fly on to the Arctic Circle before returning to Africa, though it is too far for a single butterfly to travel. Up to six generations, larva-caterpillar-butterfly may pass on during the trip.

Human annual migration for summer and winter pastures for livestock and food seems to have ended about 1,300 years ago. European nomadic tribes moved north and south across mainland Europe during the first 700 years of the first millennium.[8] Would those travelling people have fared better on their way if they could sense the earth's magnetic field and polarised sunlight? That humans lack these senses is the probable reason why we find bird migration so enigmatic. Despite the end of annual food journeys, the human quest for nourishment still caused what is surely the biggest ever human movement

---

7. For further details, see 'Painted Lady migration secrets revealed' on the Butterfly Conservation environmental group website at **butterfly-conservation.org/news-and-blog/painted-lady-migration-secrets-revealed**
8. *The Outline of History*, H. G. Wells, Waverley Book Co. Ltd., 1925, p314.

– some 300 million Bantu-speaking people of central Africa travelled east and south over several hundred years from about 1000 BCE to settle in those wider African parts. They had grown in numbers, worn out their resources and needed fresh lands to grow food. Yet still, human migration is widespread in current times, though not in a regular search for food – economic prosperity, escape from wars, and forced migration such as the African slave trade, are among the reasons.

In the modern age, the nearest humans come to an annual airborne migration is the summer holiday! It can be argued though that this is as

Redstart. (John McCue)

beneficial to our well-being as the birds' annual search for food and salubrious surroundings. Notwithstanding refreshing vacations, the eventual downside of our flights around the world will be contamination of the atmosphere leading to a degradation of the environment around our planet. Will this warning be heeded?

Human culture has been baffled by bird migration for many hundreds of years. Born in northern Greece, Aristotle studied bird behaviour and concluded that the disappearance of the summer redstarts, and the appearance of winter Robins, could be explained by the transformation of the former into the latter and back again.[9] Well known also is the medieval theory that when Barnacle Geese (which breed in the northern Europe summer and arrive in the UK during the winter) appear, they have just emerged from barnacles in the sea.

Summing up, birds can use the sun, stars and earth's magnetic field as guides on their journeys, and sometimes use more than one of these tricks at the same time. Birds also follow landmarks on the ground, of course. We know though that built-in clocks and compasses will help young birds home if they get left behind!

---

9. 'Ancient Explanations of Bird Migration', Richard Armstrong, courtesy of the University of Houston, **uh.edu/engines/epi2228.htm**

Starry, Starry Night

A History of Astronomy Sketching and Art

Mary McIntyre

Our early ancestors have been studying and painting the night sky for millennia. In Lascaux, France there are 17,000 year old cave paintings that were discovered in 1940, and in 2000 Dr Michael Rappenglueck of the University of Munich discovered that two of the paintings depict the stars in the area around Taurus and the Summer Triangle. He also discovered a 14,000 year old cave painting that depicted the constellation Corona Borealis in the Cueva del Castillo (Cave of the Castle) in the mountains of Pico del Castillo, Spain. As of 2022, these remain the earliest evidence of human interest in the night sky. A petroglyph rock carving depicting constellations was created about 5,000 years ago in Arizona, another in Phoenix recording the supernova in Scorpius in 1006.

The first astronomers didn't have access to camera equipment so they drew or painted what they saw, but it wasn't just astronomers who were fascinated with the heavens; astronomical apparitions have inspired artists throughout our entire history. Here is a look at some of those works of art.

The Sun, Moon and Eclipses

Many civilisations worshipped the Sun, so it appeared extensively in art and sculpture. The Aztec Sun Stone, one of the most famous pieces of Mexican sculpture, was probably carved between 1502 and 1520 and is a circular stone monolith with a diameter of 358 centimetres (141 inches). It weighs almost 25 tonnes and has concentric carved rings on the surface with the face of the solar deity at the centre. In European art, the Sun was often depicted in black, red and white, colours that corresponded to alchemical changes.

Early artistic representations of celestial bodies and atmospheric effects often had religious or allegorical significance and were frequently painted with faces because they were representing deities. One gorgeous example of this is "Sole Luna stelle e venti" (Sun Moon Stars and Winds), an illustration from the 1230 treatise *Sphera vulgare novamente* by Joannes de SacroBosco.

The Aztec Sun Stone or Calendar Stone. This 358 centimetre diameter stone was probably carved between 1502 and 1520 and the face at the centre depicts a solar deity. It is displayed in the Aztec (Mexica) Gallery, INAH, National Museum of Anthropology, Mexico City. (Wikimedia Commons/Gary Todd/CC0 1.0 Public Domain Dedication)

"Sole Luna stelle e venti" (Sun Moon Stars and Winds) – an illustration from the 1230 treatise *Sphera vulgare novamente* by Johannes de SacroBosco. The Sun, Moon and winds are all depicted with faces. (Wikimedia Commons/Mauro fiorentino Phonasco et Philopanareto)

Philosophers also created charts that represented models of the universe. A stunning chart representing Copernicus's heliocentric universe was produced by the Dutch–German cartographer and cosmographer Andreas Cellarius (c.1596–1665) in 1661. Earth and the planets are in orbit around the Sun, with zodiac signs along the ecliptic.

Heliocentric universe model chart by Andreas Cellarius, 1661. This beautiful illustration places the Sun at the centre of the universe, a model proposed by Nicolaus Copernicus. (Wikimedia Commons/Andreas Cellarius)

During the Medieval period, art was also used for practical purposes in the form of illustrated calendars. One example produced for farmers was the *Faltkalender mit Monatsbildern* (folding calendar with monthly pictures) c.1400. Each month had the zodiac sign, the hours of daylight and an illustration showing agricultural tasks for that month.

The Moon has featured extensively in art throughout every period of history. Almost every artist has painted nocturnes at some point in their career, and even if the Moon itself is not present, the moonlit cloudscapes and foregrounds make the paintings incredibly beautiful. The oldest known map of the Moon was created from naked eye observations by the English physicist and natural

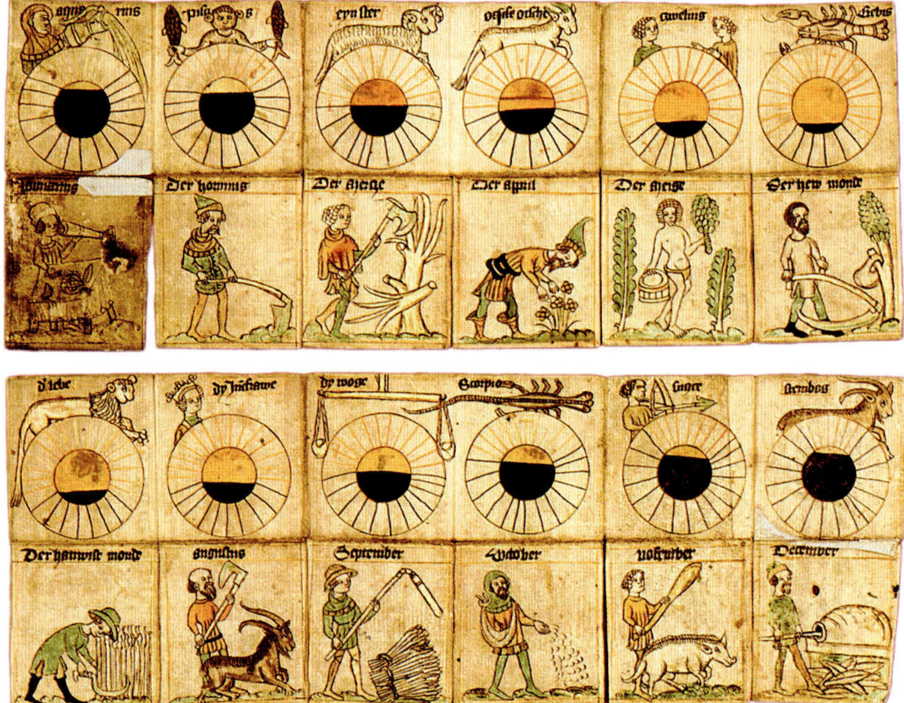

Faltkalender mit Monatsbildern (Folding calendar with monthly pictures). This calendar was an aid for farmers. The hours of daylight were shown by different amounts of black obscuring the Sun's disc, and the drawings showed which agricultural tasks should be completed each month. (Wikimedia Commons/Unknown author/Berlin State Library/Prussian Cultural Heritage/SB-PKK, Lib. pic. A 92)

philosopher William Gilbert in around 1600, and was published in *De Mundo nostro Sublunari Philosophia Nova*, 1651. Unlike most of his contemporaries, Gilbert believed the lighter parts of the lunar surface were water.

The invention of the telescope in the early-1600s lead to an entirely new era of discovery as astronomers sketched or engraved what they saw at the eyepiece. The first astronomer to create an eyepiece sketch was the English astronomer and mathematician Thomas Harriot (c.1560–1621), who made a drawing of the Moon on 26 July 1609 as it looked through his recently built "Dutch Trunke" telescope. Because this telescope did not have as good resolution as Galileo's telescope, Harriot's sketch does not have as much crater detail as those seen in Galileo's sketches, although it was the first drawing of the Moon to show craters. He made further observations and produced a lunar surface map. Many astronomers today still continue to create stunning eyepiece sketches, because translating what the eye sees and putting it onto paper makes them feel a real connection with the object they are drawing.

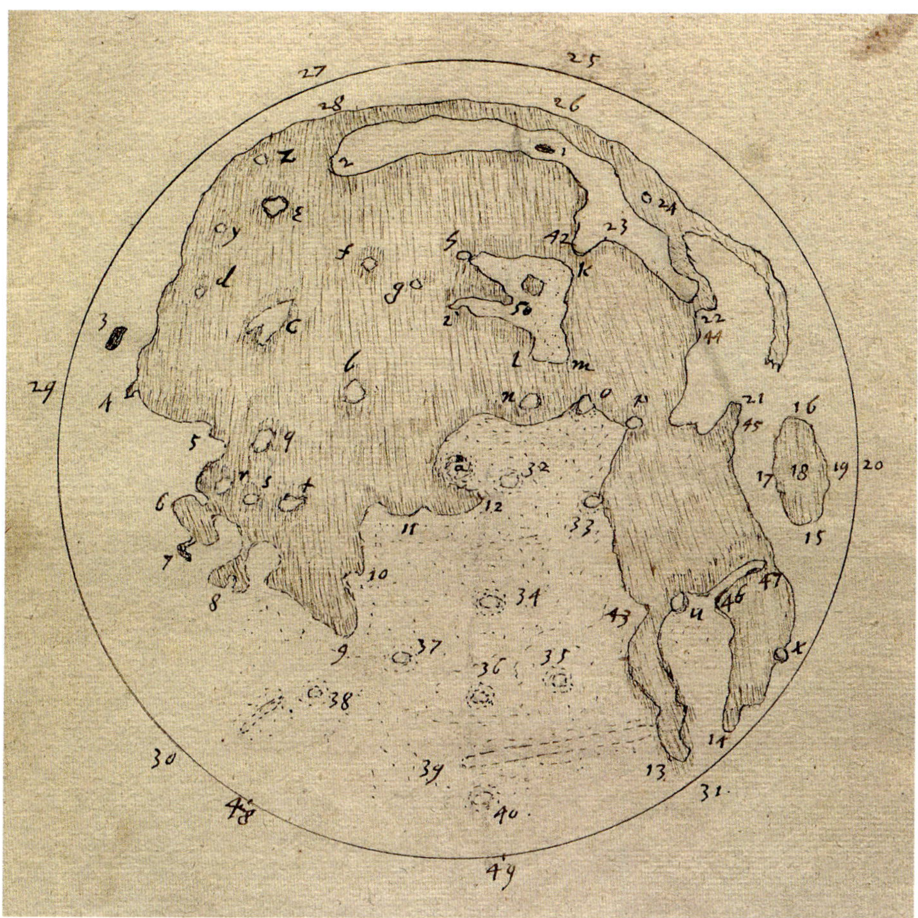

Drawing of the Moon by Thomas Harriot. This sketch was created on 26 July 1609, thereby pre-dating Galileo's first telescopic Moon drawing by four months. Although his "Dutch Trunke" telescope did not have great resolution, Harriot's lunar drawing was the first to feature lunar craters. (Wikimedia Commons/Thomas Harriot)

The Moon played an enormous role in mythology and cultural symbolism in Japanese and South East Asian culture. *One Hundred Aspects of the Moon*, a series of woodblock prints created by the Japanese artist and printmaker Tsukioka Yoshitoshi between 1885 and 1892, were a celebration of this symbolism.

In 1902, the French illusionist and film director Georges Méliès (1861–1938) produced *Le Voyage dans la lune* (A Trip to the Moon), a silent film considered by many to be the first sci-fi movie. The scene where the spaceship crashes into the Moon's eye is one of the most iconic in film history.

In 2016, installation artist Luke Jerram launched the breathtaking travelling art installation Museum of the Moon; a 7-metre diameter inflated model of

"The Moon on Musashi Plain" (Musashino no tsuki) by Japanese artist Tsukioka Yoshitoshi is a woodblock print from the series 'One hundred aspects of the moon (Tsuki no hyakushi)', 1892. The Moon played a huge role in mythology and symbolism in Japanese culture, and this series contains 100 different paintings each celebrating the Moon. (Wikimedia Commons/Tsukioka Yoshitoshi/Art Gallery of South Australia/Accession Number 20054G73)

Screenshot from *Le Voyage dans la lune* (A Trip to the Moon) (1902) a film by Georges Méliès, released on 4 October. In this iconic movie scene, the spaceship hits the Moon in the eye. (Wikipedia/ Georges Méliès/Roger-Viollet)

the Moon. The incredibly detailed surface was created from NASA imagery, the sphere being gently illuminated from within. The installation has toured extensively since it launched, attracting many thousands of visitors. It really is an amazing experience to spend time standing so close to such a huge and detailed model.

Museum of the Moon at Birkenhead on 29 October 2019. First launched in 2016, this travelling art installation is a 7-metre diameter model of the Moon with a detailed surface created from high resolution NASA imagery. It has attracted thousands of visitors in numerous amazing locations. (Wikimedia Commons/Phil Nash from Wikimedia Commons/CC BY-SA 4.0 & GFDL)

Eclipses, along with many other apparitions or atmospheric phenomena that could not be fully explained, were seen as bad omens and were surrounded in superstition. Because of this they feature extensively in art and were depicted in many illuminated manuscripts from the medieval period. The 1552 *Augsburger Wunderzeichenbuch* (Augsburg Book of Miracles) was a bound collection of 167 watercolour and gouache paintings of "miraculous" events; 60 of these were normal astronomical events, many of which had been incorrectly attributed to being the cause of some kind of natural disaster.

On 12 April 1009 there was total lunar eclipse, but a couple of weeks prior to this on 29 March a partial solar eclipse occurred which was visible from across northern Europe; this one was one of three partial solar eclipses during 1009, although the other two were not visible from Europe. From Germany, the 29 March eclipse had the Moon obscuring about 40% of the Sun, which may have led to a perceived reduction in sunlight. Folio 35 in *Augsburger Wunderzeichenbuch* is a depiction of both the solar and lunar eclipses from 1009 in addition to a "huge burning torch" falling from the sky with a "crashing noise". This could have been a fireball meteor with associated sonic boom.

Folio 35 from the *Augsburger Wunderzeichenbuch* depicting lunar and solar eclipses and probably a bright fireball meteor that all occurred in 1009. The caption reads, "In the year A.D. 1009 the sun went dark and the moon was seen all blood-red and a great earthquake struck and there fell from the sky with a loud and crashing noise a huge burning torch like a column or a tower. This was followed by the death of many people and famine throughout Germany and Italy. More people died than remained alive". (Wikimedia Commons/Anonymous Author/Jessie Huffaker)

The oil on panel painting "Astronomers Studying an Eclipse" is attributed to the French painter and illustrator Antoine Caron, 1571. The original title is not known. Alternative titles given to this painting include "Astronomers watching a solar eclipse" and "Astronomers (or Astrologers) Studying an Eclipse". (Wikimedia Commons/Antoine Caron)

These three bad omens from the sky were linked to an earthquake – there was an earthquake reported in Lisbon that year – and famine throughout Germany and Italy. Another folio in the same book depicts another event in 1362 which involved multiple bad omens in the form of a large stone falling from the sky in heavy wind and rain (again, probably a bright meteor) followed by a solar eclipse.

In 1570 a painting that is attributed to the French Renaissance court painter Antoine Caron depicts a solar eclipse over Paris. The original title of this painting is not known, but is has since had a couple of name changes! In 1947 it was given the title "Astronomers Studying an Eclipse". It is now known by some, probably erroneously, as "Dionysius the Areopagite Converting the Pagan Philosophers".

Stars and the Night Sky

Indigenous peoples across the world divided the night sky into constellations and created celestial maps. The star patterns represented characters or items that were important to local culture; each had their own constellation mythologies and artists created stunning artworks of those stories. The term "zodiac" has become synonymous with astrology and horoscopes in the modern day, but the true definition refers to the band of sky near to the ecliptic; i.e. the path that the Sun, Moon and planets all follow. The earliest known classical depiction of the zodiac dates back to around 50 BCE and is a relief sculpture that was on the ceiling of the Hathor Temple in Dendera. This is now on display at the

"Heavenly Body in the Night Sky" by Albrecht Dürer, oil on panel about 1469 which was on the reverse of his painting of Saint Jerome. This abstract painting was very different from his usual realism style and leaves the viewer feeling uneasy. (Wikimedia Commons/ Albrecht Dürer/National Gallery/ Accession Number NG6563)

Louvre in Paris. Almost every culture has a plethora of art depicting their own zodiac and mythology characters. The Greek/Roman constellations we know today were not named until around the fourth century BCE.

Stars feature heavily in nightscape paintings and they usually evoke feelings of calmness and tranquillity. One piece of art that leaves you feeling more unsettled than calm is "A Heavenly Body in the Night Sky" painted by the German artist Albrecht Dürer around 1469. It is an abstract depiction of a celestial body that is pulsing with light. This stunning work differed from his usual realism style, but as Dürer wrote very little about it, it is not known if this object is a bright star, a comet or a meteor. This was not the only time that celestial bodies appeared in Dürer's work; the sky in the 1514 engraving "Melencolia I" features an object that could be rainbow, a comet or a planet.

In the late-1800s, some of the best known abstract night sky paintings were produced by Vincent Van Gogh. In 1888, he painted "Café Terrace at Night" and "Starry Night Over the Rhône". The former was the first time Van Gogh had painted a starry sky in his works, although the main focus is the artificially lit café. The latter featured Arles at night, painted from

"Starry Night Over the Rhône" by Vincent Van Gogh. This was the view of Arles at night as seen from the banks of the Rhône, where he could capture the reflections of the lights in the water. The painting also features The Plough, although it was never visible in that part of the sky. (Wikimedia Commons/Vincent Van Gogh/Musée d'Orsay/Accession Number 78696)

the bank of the Rhône from a position where he could see and paint the reflections of the gas lights in Arles, whose bright colours contrast to the paler coloured stars above. It features the Plough, an asterism in Ursa Major (the Great Bear), although this asterism is never visible in this part of the sky from the scene's viewpoint.

In 1889 Van Gogh painted "The Starry Night". Although abstract, the foreground depicts some of the view he had from his hospital room at Saint-Rémy-de-Provence, where he was receiving inpatient treatment for his mental health issues. This painting features a stylised waning crescent Moon (even though this was not the lunar phase at the time of painting) as well as a bright star-like Venus and the constellation Aries (the Ram), along with a very distinctive swirling pattern thought to be inspired by Lord Rosse's 1845 sketch of the Whirlpool Galaxy. This masterpiece has become one of the most recognisable Western art paintings. Many modern art painters were inspired by the night sky, with Pablo Picasso, Joan Miró, Georgia O'Keeffe and many others creating abstract paintings featuring stars.

"The Starry Night" by Vincent Van Gogh. This painting depicts the view from Van Gogh's hospital window. Although the colour choices in this painting are similar to his other nocturnes, the swirling pattern in the sky is thought to have been inspired by William Parson's sketch of Messier 51, the Whirlpool Galaxy. (Wikipedia/Vincent Van Gogh/Museum of Modern Art/ Accession Number 472.1941/Google Art Project)

Comets and Meteors

Along with many other astronomical events, comets and meteors were seen as bad omens with comets even being labelled as "harbingers of doom", so there's no shortage of comets in art. One of the earliest depictions of a comet was of the 1066 apparition of Halley's Comet (modern designation 1P/Halley) in the Bayeux Tapestry. The comet was described as being four times the size of Venus and with a magnitude similar to a quarter of the Moon. This apparition was linked to the death of Harold II during the Battle of Hastings.

The Star of Bethlehem was depicted as a comet in the 1305 painting "Adoration of the Magi" by Giotto di Bondone but there's no evidence that a bright comet was visible at that time. The Augsburger Wunderzeichenbuch contains 30 illustrations of comets. A bright comet visible in 1007 is depicted in Folio 34; the painting is a dramatic, highly stylised representation of a comet which ties in with the belief that comets were a bad omen. Published in 1587, *Kometenbuch* (The Comet Book) – the author and illustrator of which are unknown – was based on an anonymous Spanish treatise about the significance of comets, and contains 13 watercolour paintings of comets and meteors.

The Great Comet of 1811 (modern designation C/1811 F1) had a big impact on astronomers and non-astronomers alike due to it being visible for around

Comet drawing circa 1552, unknown author, active near Augsburg, Folio 34 of the Augsburger Wunderzeichenbuch, in the "Miraculous Signs" section. The caption reads, "In 1007 A.D., a wondrous comet appeared. It gave off fire and flames in every direction. It was seen in Germany and Welschland and it fell onto the earth". (Wikimedia Commons/Unknown Author)

A painting of the sungrazing Great Comet of 1843 as seen from Tasmania by the English Australian artist Mary Morton Allport. (Wikimedia Commons/Mary Morton Allport)

Vieux Cognac bottle from 1811. The Great Comet of 1811 was visible for about 260 days, and was thought to be responsible for the long, hot summer that led to a particularly good year for wine. Merchants were selling "Comet Wine" at inflated prices for several years after, in bottles that had an embossed seal that featured a comet. (Mary McIntyre)

260 days. The comet was painted by John Linnell, William Blake and many more. The year 1811 turned out to be a particularly good one for wine and cognac, and merchants were selling "Comet Wine" at inflated prices for several years after the apparition, in bottles that contained an embossed seal featuring a comet on the front.

In 1843 a Kreutz sungrazer comet, with the official designation C/1843 D1, became known as the Great Comet of 1843. First observed on 27 February 1843 – when it was seen in broad daylight just one degree from the Sun – it was best viewed from the southern hemisphere, remaining visible until 19 April 1843. The object was depicted in paintings by English-born Australian artist Mary Morton Allport (1806–1895), who observed it from Tasmania, and by Italian-born English astronomer Charles Piazzi Smythe. The Great Comet of 1843 also inspired the Mexican composer Luis Baca to produce a piano waltz called *El cometa de 1843*.

On 1 October 1858, Donati's Comet was depicted by the English landscape painter James Poole. The most observed comet of the nineteenth century, Donati's Comet inspired many artists. In stark contrast to the 1007 painting from the *Augsburger Wunderzeichenbuch*, Poole's painting is a realistic impression of a comet; he captured the reflections on the still water in the foreground beautifully to produce a serene scene that's the antithesis of bad omens.

Meteors and fireballs were also viewed with superstition so they too appeared extensively in art. The *Kometenbuch* featured a meteor shower illustration which shows the radiant, and demonstrates a good understanding of meteor showers. A woodcut illustration of a fireball and ten kilogramme meteorite fall at 5pm on 9 April 1628 in Hatford, Berkshire features a celestial army sporting huge cannons (the source of the sonic boom!).

Donati's Comet, modern designation C/1858 L1 (Donati), painting by James Poole, 1 October 1858, oil on canvas. In stark contrast to the painting of the 1007 AD comet, this beautiful and serene painting is a more realistic view. (Wikimedia Commons/Sheffield Galleries and Museums Trust/Accession Number K1908.66)

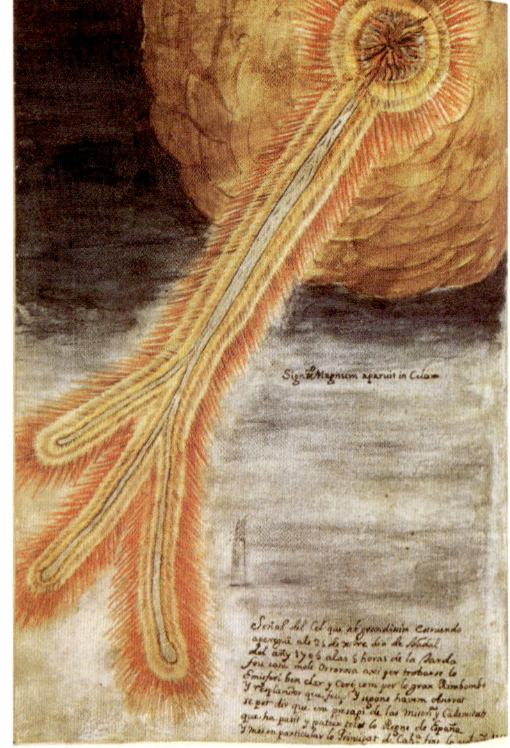

Meteor over Catalonia, originally painted by Josep Bolló, eighteenth century. This is a highly stylized painting of a bright fireball meteor visible on Christmas Day 1704 from Catalonia. This fireball resulted in a meteorite fall, following which around a kilogramme of chondrites were later recovered from the area. (Wikimedia Commons/Xavier Caballe/CC BY-SA 2.0)

At 5pm on Christmas Day 1704 there was a bright fireball resulting in a one kilogramme meteorite fall over Catalonia. This event was included in Josep Bolló's *Scientific Miscellany* and was blamed for the "miseries and calamities" that Spain and Catalonia were suffering from at that time. This painting is extremely stylised and doesn't really bare any resemblance to an actual fireball meteor.

In 1833 the Leonid Meteor Shower was said to have produced a spectacular meteor storm. It was estimated that there was between a 100,000 and 240,000 meteors per hour over a nine hour period visible from North America. People said they believed this was "a literal fulfilment of the word of God" and a sign of the imminent second coming of Jesus. There are several artworks that depict this event and it truly must have been an incredible sight to behold!

A rare, bright Earth-grazing meteor precession event was witnessed from the USA on 20 July 1860. It was observed and subsequently painted by Frederic Edwin Church, a Hudson River school painter. In contrast to Josep Bolló's work, Church's painting is photorealistic. The colour palette is reminiscent of Van Gogh's paintings, and the reflection of the fireball in the water is truly stunning.

"The Meteor of 1860" by the American landscape painter Frederic Edwin Church, oil on canvas. This unique event was reported as a rare earth-grazing meteor precession. (Wikimedia Commons/Frederic Edwin Church/Judith Filenbaum Hernstadt)

Atmospheric Optics

Phenomena such as ice haloes, parhelia (sundogs/mock suns) and arcs caused by refraction of sunlight through ice crystals in cirrostratus clouds have been studied by scholars since Aristotle wrote a treatise called *Meteorology* in around 360 BCE in which he noted his observations of parhelia along with his thoughts on how they are formed. Despite scholars having a decent understanding of the underlying mechanisms, the appearance of atmospheric optics was met with superstition and was thought to be a bad omen. One notable apparition of parhelia was on the morning of the Battle of Mortimer's Cross on 2 February 1461, when Edward of York described seeing "three glorious suns, each a perfect sun". His troops were terrified it was a sign of their imminent death, but he convinced them it was in fact a good omen. This event was re-enacted in Shakespeare's Henry VI, Part 3, with the line, "Dazzle mine eyes or do I see three suns?"

The earliest pictorial record of parhelia is in the *Nuremburg Chronicle*, an illustrated bible published in 1493, where they are depicted as the Holy Trinity. The Augsburger Wunderzeichenbuch contains many paintings of atmospheric optics and they have appeared in numerous artworks and single page broadsheets.

The most famous atmospheric optics painting is the "Vädersolstavlan", colloquially known as "The Sundog Painting". It was created in 1636 by Jacob Heinrich Elbfas and is an accurate copy of a 1535 painting by Urban Målar depicting a spectacular multiple halo display seen from Stockholm on 20th April 1535. The interpretation is unusual because the display was visible in the morning but the city shadows are from the evening. The position and orientation of the haloes and arcs relative to the Sun is also incorrect, but it is still a stunning piece of art.

Aurorae

One astronomical phenomenon on everybody's bucket list is Aurora Borealis / Aurora Australis. Most civilisations told mythological stories about it and meteorologists went on arctic expeditions to study it in more detail. The beauty of aurora has been captured in many works of art, one such work being painted in 1865 by Frederic Edwin Church. The painting was based on sketches created by Arctic explorer and pupil of Church Isaac Hayes; Hayes' ship appears in the foreground of the painting. The sky is alight with curtains of various colours of aurora and interestingly it depicts STEVE (Strong Thermal Emission Velocity Enhancement), an aurora phenomenon that was not officially investigated and named until 2016.

Sundog illustrations from the *Nuremberg Chronicle* by the German historian and cartographer Hartmann Schedel (1440–1514). In this illustrated bible the Sun and parhelia (also known as sundogs or mock suns) are depicted as the Holy Trinity. As was common at the time, each are depicted with faces. (Wikimedia Commons/Hartmann Schedel)

"Vädersolstavlan" also known as "The Sundog Painting" by the Baltic German portrait painter Jacob Heinrich Elbfas, 1636. This depicts an amazing multiple halo display seen from Stockholm on 20 April 1535. The orientation and position of the haloes and arcs have been changed so this is not how the display would have looked in person. The painting seen here is an accurate copy of "Vädersolstavlan", a 1535 painting by Urban målare, and it was restored in 1998. (Wikimedia Commons/Jacob Heinrich Elbfas/Mats Halldin/Storkyrkan)

"Aurora Borealis" by Frederic Edwin Church 1865, oil on canvas. This painting was based on two sketches done by other artists; one was by a pupil of Church, the other was by Isaac Hayes, an Arctic explorer whose ship appears in the foreground. (Wikimedia Commons/Frederic Edwin Church/Smithsonian American Art Museum/Gift of Eleanor Blodgett/Accession Number 1911.4.1)

Space Art

In the 1940s, French illustrator and astronomer Lucian Rudaux and American painter Chesley Bonestell pioneered the new genre of "Space Art" with their stunning paintings depicting the landscapes of other planets and moons. Early examples were produced long before any planetary flybys had taken place so the depicted scenes were often incorrect, but this does not detract from their beauty. One of the best known examples was Bonestell's painting of Saturn as viewed from Titan, which was published in *Time* magazine. In recent times, artists such as David A Hardy continue to produce stunning space landscape paintings using traditional and digital art techniques.

Oil paint is often the preferred medium for many space artists because the colours blend effortlessly into each other to create realistic skyscapes, and Cathrin Machin and Vicky Hunt (Vicky Paints Space) are two artists who produce stunning photorealistic oil paintings.

Like many space artists, Vicky Hunt takes her inspiration from the beauty of the cosmos, from our closest neighbour the Moon to distant galaxies. She creates paintings of real astronomical objects as well as imaginary free-flowing nebulae and skyscapes. She spreads her love of space through art and

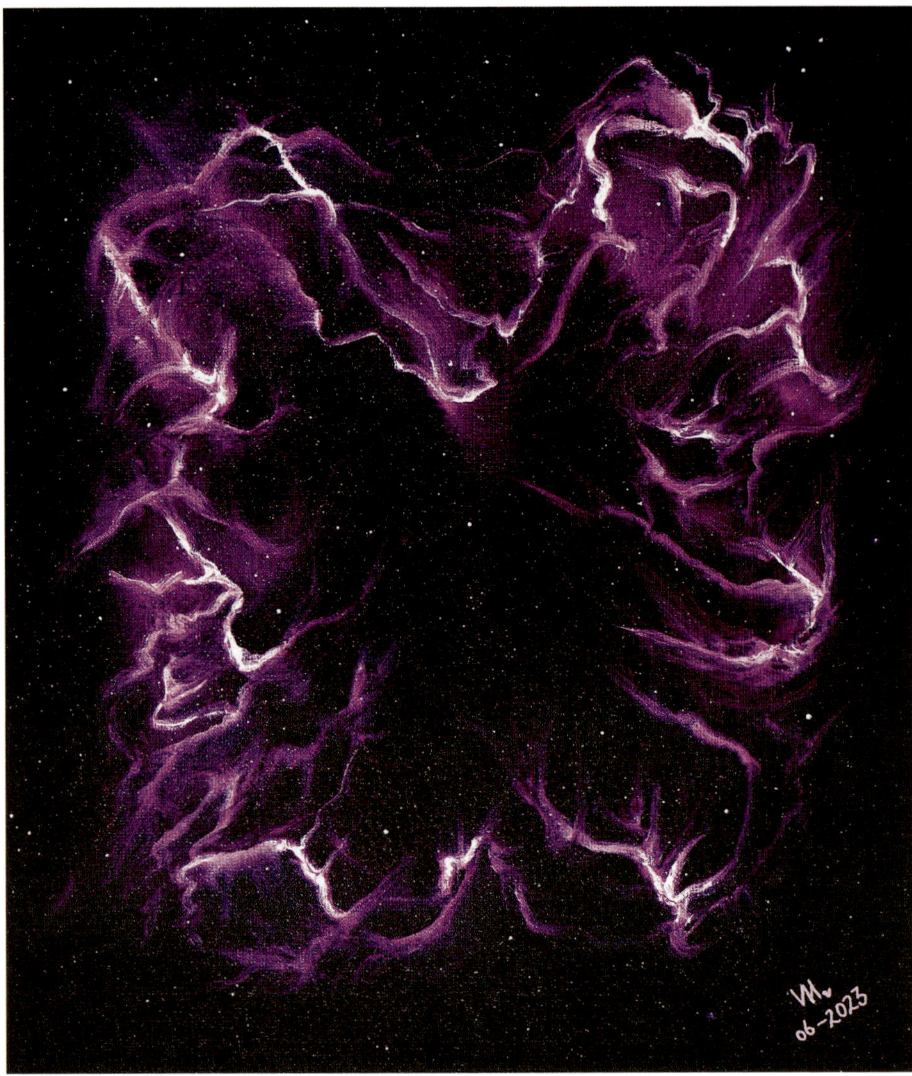

"Cosmic Love" by Vicky Hunt, oil on canvas board. This free-flowing nebula-inspired painting was created by applying a layer of transparent magenta on the canvas, then picking up the colour by adding titanium white. The blending process allows the pink to become opaque and visible, creating this nebula from imagination. The slight resemblance of a heart is there to remind everyone that the universe is always there for us. (Vicky Hunt/VH Space Art)

encourages people to look up and see the endless possibilities that the sky holds. She even infuses her paintings with genuine cosmic dust to really bring the universe into people's lives.

Many artists within the acrylic paint pouring community such as Olga Soby, Molly Leach (Molly's Artistry) and the author of this chapter have used fluid acrylic techniques to create abstract pieces that are inspired by space.

"Ice World" by Mary McIntyre, fluid acrylics on canvas. This abstract painting was inspired by a David A. Hardy painting of an ice cave on Pluto, created many years before any probes had visited the planet. Using paint diluted with a pouring medium and water, the outside of the painting was created using a hairdryer. Once dry, the rest of the painting was finished with brushes. (Mary McIntyre)

Conclusion

Imagery from every period of astronomy and space history still continues to inspire artists and designers the world over, and in turn paintings depicting space and astronomy inspire more people to take an interest in space. If you look at paintings, digital art, textiles, clothing or ceramics, space is absolutely everywhere and thanks to the *James Webb Space Telescope*, *Artemis* and Space Exploration Technologies Corp. (usually referred to as SpaceX), public interest in space is at its highest since the *Apollo* missions. Huge advances in telescope and imaging technology have revolutionized photographic and artistic views of our universe. Even though we see amazing new space images daily and, as a society, we have a much better understanding of the universe around us, products inspired by early celestial art that depict the Sun and Moon as deities with faces, still remain extremely popular in home décor, jewellery and clothing. It's amazing to think that those very first artists who were trying to understand and depict the universe are still inspiring art all these years later.

Ceramics exhibited in a street of Deruta depicting the Sun and Moon and faces by Bultro. Although we no longer view celestial bodies as deities, it is still extremely popular to see the Sun and Moon with faces in fabric designs, home décor and ceramics. (Wikimedia Commons/ Bultro/Creative Commons Licence CC BY-SA 4.0)

Time, Space and Stars

How Celestial Clocks Connect Us to the Universe

Robert Massey

Human beings, like other animals, are hardwired to notice and respond to the passage of time. Most of us plan our work and leisure around day and night, and even in our light saturated world we tend to curtail our outside activity after dark. It also turns out that many of the clocks and navigation systems we rely on only work because of our deeper understanding of how time is embedded in our universe.

The Sun is the most obvious marker of time as the Earth turns. Unless you live in the northernmost regions of the world and experience continuous night

The stone circle at Stonehenge in Wiltshire, England. (Wikimedia Commons/Gareth Wiscombe)

for weeks or months in the winter, our nearest star appears to rise in the east and sets in the west. At night the stars move in a similar way, rotating around the poles of the sky. And over longer periods the changing position and phase of the Moon from one night to the next, and the variation in the length of daylight and night define months and the course of the seasons.

As early as 40,000 years ago, our ancestors may have recorded the cycle of lunar phases, as the Moon appeared to grow from a crescent to a fully illuminated disc, and then shrank down again. The famous cave paintings in Lascaux in southern France include an intriguing series of 29 marks interpreted as tracking our natural satellite over a month.

Built thousands of years later, the Neolithic monuments predominantly but not exclusively found in north-western Europe (the oldest stone circle may actually be at Nabta Playa in Egypt and there are many later examples in Mesoamerica) have long been recognised for their connection to the sky. Sites across France, Ireland and the UK have a deep celestial connection.

On the largest scales, the stone circles of Stonehenge and Avebury in Wiltshire, and Callanish on the Scottish Isle of Lewis were constructed from around 5,000 years ago. Their stones mark the rising of the Sun at midwinter and midsummer, and the extremities of the rising and setting positions of the Moon. Whatever their function, these structures were not intended to note the passage of time within a single day.

It is also hard to believe that Neolithic peoples needed megalithic monuments to find the time of year, given the obvious changes in the weather and the height of the Sun in the sky. But it does demonstrate an awareness of the sky, and a precision in indicating significant points in the calendar, acknowledged to this day by people who gather at places like Stonehenge to watch the sunrise at the summer solstice.

More personalised and certainly far smaller, sundials use a projecting component, a gnomon, to show how time passes between sunrise and sunset, with the shadow from the gnomon cast on to a dial plate. The shadow then moves as the Earth turns and the Sun appears to move across the sky.

The first sundials were developed around 3,500 years ago in ancient Egypt, and in the earliest example the length of the shadow indicated the time of day. The technology slowly spread across the ancient world, with early sundials in China in around 800 BCE, in Babylon by 300 BCE, while in Classical and Hellenistic Greece (510 to 31 BCE) and the Roman Republic they became ubiquitous, with the Roman comic playwright Plautus complaining how they even dictated when to eat. Some of their remains survive to this day, such as the Tower of the Winds in Athens and Interamna Lirenas in Italy.

A sundial in Kew Gardens, London. (Wikimedia Commons/Matěj Baťha)

The duration of daylight varies with the seasons and despite this the Greeks and Romans divided both daylight and night into 12 hours each. These 'unequal hours' change dramatically from summer to winter, with the variation more pronounced the further north the sundial is sited.

Dial plates on more advanced sundials can normalise time to a system of 24 equal hours (at least when the Sun is above the horizon), pointing the gnomon towards the celestial pole, and in the modern world can also take into account the Equation of Time, which describes how the apparent position of the Sun, and so solar time, varies because of the shape of the orbit of the Earth. With this in place and allowing for corrections associated with time zones, they can be used to tell the time to an accuracy of about a minute.

Arguably some of the most advanced devices connected with time before clocks, astrolabes are another innovation originating in ancient Greece, with their invention attributed to the mathematicians and astronomers Apollonius of Perga and Eudoxus of Cnidus. They have a large number of different functions, and fundamentally link time to the position of the Sun and stars.

The most advanced astrolabes include an outer ring marking angle or sometimes hours; a plate made for a specific latitude, which has coordinates of the sky and the horizon around the edge; a rotating disk overlaying the

A fragment (Fragment A) from the Antikythera mechanism. (Wikimedia Commons/Marsyas/ GNU Free Documentation License, Version 1.2)

plate with holes for the brightest stars, including a ring that marks the ecliptic, the apparent path of the Sun across the sky, which may be divided into constellations or months; a bar sitting across the astrolabe used to measure the angle of objects in the sky; and the alidade, a sight on the back of the device.

These complex instruments can be used to measure the position of the Sun, stars and planets, and to find local time. Astrolabes were developed further in the Islamic world, where they were important for determining prayer times, and to find the direction of Mecca, both of which are vital for observant Muslims.

The ancient world also saw the development of mechanical devices depicting astronomical events. The archetypal example is the beautiful Antikythera mechanism, discovered by Greek fishermen in 1901, which was constructed around 2,200 years ago and is now on display in the National Archaeological Museum in Athens.

For more than a century its precise purpose was unknown, but in the 2000s a team led by Professor Mike Edmunds at the University of Cardiff used X-rays and surface scanning to analyse its interior and inscriptions. They deduced that its complex system of 37 gears simulated the movements of the Sun and Moon through the constellations of the zodiac, and could be used to predict lunar and solar eclipses.

As much as sundials, mechanical clocks are a physical recognition of the connection between human activity and the wider cosmos. The movement of the hands of the clock mimics the path of the shadow of the Sun, and almost all clocks in use in civil society are aligned to the mean solar day.

Astronomical clocks go further, indicating the positions of the Sun, Moon, constellations and sometimes planets. One of the first of these, actually a large clock tower powered by a water wheel, was developed by the Chinese polymath Su Song in 1088. His tower stood 12 metres high, and included a rotating armillary sphere (where rings designate the lines of latitude and longitude in the sky) and a celestial sphere designating the constellations. The clock predated the earliest European examples of mechanical astronomical clocks by around 240 years, but was sadly dismantled by the invading Jurchen army in 1127, who were then unable to reassemble it. A full size working replica now stands in Kaifeng, China, where the original was built, and a 1:6 scale model was constructed by the Science Museum in London in 1965.

The same period saw the development of water clocks with astronomical features in the Islamic world. The polymath Ibn al-Razzaz al-Jazari from Upper Mesopotamia described elaborate timekeepers, including the notable example of a "castle clock" that showed the motions of the Sun and Moon, lunar phases and the zodiac.

In the fourteenth century Europeans built their first mechanical astronomical clocks. Richard of Wallingford, an English scientist and mathematician constructed a clock for St Albans Abbey, where he was the abbot, in the 1330s. In northern Italy, astronomer Giovanni de Dondi completed another example in Padua in 1364. Neither of these survives, but plans describe how they showed the Sun, Moon and the five naked eye planets.

Other examples from the same epoch are found across Europe, including in France, Germany, Sweden and the Czech Republic, where the clock on the Old Town Hall in Prague is a major tourist attraction and the oldest one still in operation. There is something compelling about watching a device built by our ancestors more than 650 years ago, albeit based on a geocentric universe. The slow movement of the Sun across the sky from day to night, its longer term position against the constellations of the Zodiac, and the slowly changing phase of the Moon contrast with the pace of our lives today, where we fret about delays of even a few seconds.

The astronomical clock on the old town hall in Prague, in the Czech Republic. (Wikimedia Commons/Steve Collis/CCA 2.0 Generic license)

Less helpful for timekeeping, but elegantly portraying another aspect of the nearby universe, orreries model the movement of the planets in the solar system. Named after the 4th Earl of Orrery, Charles Boyle, they are a mechanical device for demonstrating the relative speeds of the planets around the Sun, and sometimes their moons. Spheres in orreries are often scaled to the true sizes of the planets, but not compared with the distances between them, as this would make them impractically large.

Orreries were mostly developed after the Copernican revolution where the Earth lost its imagined place at the centre of the universe, replaced by the Sun, before the solar system itself was understood to be one of a huge number of planetary systems. Like the Antikythera mechanism they use gearing to model the different orbits (all approximated to circles) and clockwork to set them moving. They are related to planetaria, sometimes described as incorporating projection orreries, as an attempt to create a visual impression of the cosmos. The archetypal orrery, created and named for Charles Boyle, the 4th Earl of Orrery, was made by the clockmaker John Rowley in 1712/13 and is now in the Science Museum in London.

This untitled painting by English artist and portrait painter Joseph Wright of Derby, known as "A Philosopher Giving that Lecture on the Orrery, in which a Lamp is put in place of the Sun or The Orrery", is from the collection of Derby Museum and Art Gallery. Orreries are physical models of the solar system and typically depict its Sun, planets and moons. (Wikimedia Commons/Joseph Wright/Derby Museum and Art Gallery/europeana.eu)

have crossed the Atlantic, what in the UK is known as a 'Mars Bar' is called a 'Milky Way' in the USA, and the British 'Milky Way' is more similar to an American 'Three Musketeers'. (Experts might argue that the re-named products are not completely identical, but they are certainly similar enough to my uneducated tastes.)

Mars also produces a series of fruit flavoured chews, which were originally 'Opal Fruits' in the UK, and the functional 'Fruit Chewies' in America. These were rebranded as 'Starburst' at the height of the space-race and public interest in all things extra-terrestrial, and the name change was eventually applied in Britain too. The idea is presumably that an explosion of taste in your mouth is similar to an exploding star, a supernova!

Speaking of which, the 'Nova' was of course a highly successful model of car (made by Chevrolet in the US, and by its sister brand Vauxhall in the UK). Once again, it's possible that there is a hint of an astronomical connection. However, *nova* is just the Latin word for 'new'. This is why (super) novae are called that – they appear as 'new' stars. So, 'nova' can simply be a declaration of something original. (There's a myth that the Nova car didn't sell well in Spanish-speaking countries, as *no va* means 'it doesn't go'. It is just a myth, though, as sales of that model were entirely comparable with the Nova's success in others parts of the world.)

Similarly, the Mitsubishi Eclipse, a sporty compact car, was sold in many markets (not including Europe or the UK) for over twenty years; but although eclipses are unquestionably astronomical events, the car model was named after a famous racehorse.

Other automotive connections are clearer, though. The Ford Galaxy (and, earlier, its Galaxie) is a deliberate reference to the ideas of large and impressive size. The Vauxhall Astra's name is derived from the Latin word for a star (*aster* comes from the earlier Greek *astron*, and is present at the start of other 'astro-' words such as astronomy, astronaut, and indeed astrology). One particular type of star, a pulsar, lent its name to a Nissan design, which was known as a Datsun Cherry in many parts of the world. Quite how the same car can be a Pulsar or a Cherry is a mystery best left to the advertisers and marketers rather than engineers, and the idea of a stellar object which emits intermittent beams due to its rapid rotation does not immediately suggest a means of transport.

Constellations are often chosen as a name for something, as they have both astronomical and astrological significance. For cars, this describes three more Ford models: the Taurus, the Scorpio, and the Orion.

One step up from naming a car model is choosing the name for an entire company. Polestar is a relatively new firm, associated with Volvo in Sweden, which grew out of a racing team (now, confusingly, renamed Cyan). The Pole

An old version of the Subaru logo compared with the Pleiades. The six stars in the logo represent the individual companies which merged to form the corporation. (Allen McCloud/Clément Bucco-Lechat (Logo Photo)/ NASA, ESA, AURA/Caltech, Palomar Observatory (Pleiades Photo)/ Licence CC BY-SA-3.0)

Star itself, Polaris, has long been used as a symbol of reliability and constancy, so has excellent associations. Still small, Polestar is aiming to produce high-performance electric vehicles. That is also true of another new manufacturer, Lightyear, whose cars include multiple solar panels to top up the batteries; perhaps the first direct use of an astronomical effect in an automobile! Perhaps the suggestion of covering a long distance is part of the thought behind the name, although many people continue to believe, incorrectly, that a light year is a measure of time.

Finally, in the world of motor transport, is a much longer-established name; Subaru. This is the Japanese word for 'unity,' but it is also used as the name of the Pleiades star-cluster in the constellation Taurus. Although there are over 800 stars in the cluster, seven of them are commonly identified as the 'Seven Sisters' in Western tradition; but in Japan, the cluster is often represented with six bright stars. The Subaru Corporation was formed by the merger of six separate companies, one of them much larger than the other. This is why the manufacturer's logo is one large star, with a cluster of five smaller ones alongside it.

Given the links of astronomy to cutting-edge research and observation, it is perhaps surprising that it hasn't inspired more names in the high-tech world of computing and electronics. There is one big exception to that rule; since 2009, Samsung has used the 'Galaxy' name for its best-selling line of mobile phones, tablets, and more recently smartwatches. There has never been an official explanation of the name, but one historian of the company links it to an executive's favourite brand of wine (an expensive blended red).

In computing itself, astronomical names are very limited. Jupiter Cantab was a computer hardware company (unsurprisingly based in Cambridge); it produced a home microcomputer called the Jupiter Ace in the early 1980s, a direct competitor to the better-known products Sinclair and Acorn. The Ace used FORTH as its programming language rather than BASIC, and this sealed its fate; only 5,000 or so units were ever sold.

FORTH, as a language which requires very little memory overhead, lives on in systems which need to be small and simple, such the European Space Agency lander Philae, which was part of the Rosetta comet-landing mission in 2014. And, by coincidence, it is to comets that we now turn!

For many years, Comet was a staple of British high streets and retails parks, selling electronics and white goods. It shut its doors in 2012, but since 2019 the brand has returned as an online retailer of much the same produce range as before. The company was founded in the 1930s, recharging radio batteries for its customers. It moved into more general retail in the 1950s, and was one of the first national chains to offer a 'cheapest available' price guarantee. Its logo, of a blazing star with a trail stretching over the company's name,

A typical Comet store, at Park Road Retail Park, Pontefract, West Riding of Yorkshire, England (Wikimedia Commons/Betty Longbottom/Licence CC BY-SA 2.0)

The Gloster Meteor F.8 pictured in flight at the Centenary of Military Aviation Airshow, 2014. (Wikimedia Commons/Chris Phutully/Licence CC BY-2.0)

was instantly recognisable; but the 'star' iconography does little to dispel the ancient confusions over what comets and meteors actually are!

In fact, given that comets' distinctive tails are gases being burned off from their surfaces, and that meteors burn up in the atmosphere, it is easy to see why they have become symbols of the bright and the spectacular. However, they are not obvious names to link to stability, so it is surprising that two early and influential aeroplanes were named after them.

The British-made de Havilland Comet was the world's first jet airliner, with models in service from the 1950s to the 1980s. It had to be redesigned after several high-profile disasters; another historical association of comets is as portents of awful events. (The name was chosen as a reference to an early de Havilland aeroplane made for racing in the 1930s, where the 'bright object in the sky' name was a better fit for its garish colour scheme.)

The Gloster Meteor jet-fighter saw service at the end of the Second World War, and remained in use until the 1950s. With engines designed by the great engineer Frank Whittle, it was capable of record-breaking speeds (606 miles per hour in 1945). Unfortunately, like its namesake, the plane was liable to burn through fuel very quickly; although this rarely resulted in meteoric burn-up, it was generally considered to be an unsafe vehicle and accident statistics bear this up.

Astronomy has inspired various games over the years, although not as many as space exploration or more general science-fiction themes. There is an astronomy edition of the popular *Top Trumps* game, where players compare the values of various statistics about planets and other heavenly objects. The card game *Fluxx*, created by the American game designer Andrew Looney in 1996 and which features rules which can be modified as play progresses, also has an astronomy version. It is interesting that these are both card games; it is a format which allows for a lot of information about the particular subject being featured, and, in the case of *Fluxx*, allows an opportunity to use stunning photography.

In fact, these twentieth-century games are following in a long tradition of card games. In 1828, the London-based publisher and bookseller Charles Hodges launched his game *Astrophilogeon* which featured a beautifully illustrated deck of 60 playing cards, half of which depicted constellations and the rest showing terrestrial maps. The object of the game was for players to obtain corresponding pairs of cards, matching the constellation cards with those depicting various regions of the Earth.

The fashion for astronomical themes in the Victorian period led to some extraordinary publications. One of the most interesting was the 1824 boxed-set called *Urania's Mirror*. This consisted of some thirty large cards

The *Astrophilogeon* collection was a deck of playing cards devised by the bookseller and stationer Charles Hodges of 27 Portman Street, Portman Square, London and printed by Stopforth & Son in 1828. The initial game involved matching the celestial latitude of a constellation (top left of card) with similar latitudes for countries on Earth (half of the cards were maps of various countries). Later editions also assigned traditional card values (such as 'three of hearts') to each card, so that the deck could be used for other games. The classical gods associated with planets were used as some of the court cards. Later editions of the deck included a card for Neptune, which was not discovered until 1846. (Wikimedia Commons/Charles Hodges/Stopforth & Son)

This image shows the front cover of the box containing *Urania's Mirror*. First printed in the 1820s, the set of large (8 inch × 5.5 inch) plates printed on heavy card were accompanied by an introductory book. The novelty of having pierced holes for the stars ensured the set's enduring popularity. (Wikimedia Commons/Jehoshaphat Aspin/ *A Familiar Treatise on Astronomy*/Sidney Hall/OU Academy of the Lynx)

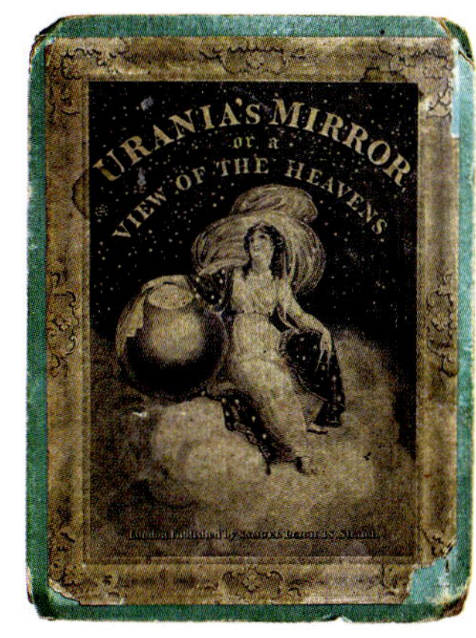

(roughly A5 size), each illustrating one or more constellations. An accompanying book was also available. What set this set aside from simple star charts was that the major stars were pierced through the card, the size of each hole representing the brightness of the star. The result was that, when the cards were held up to the light, viewers would see how the constellation appears in the night sky! Unfortunately not many complete sets remain; too many of the early purchasers held their star-cards up to candles, with predictable results.

Plate 10 in *Urania's Mirror* depicted the constellations Boötes, Canes Venatici, Coma Berenices, and Quadrans Muralis. The artist was anonymous, the work being credited only to 'a lady'. Despite this, historians now attribute the illustrations to Rev. Richard Rouse Bloxam (1765–1840), who was a teacher and under-master of Rugby School. (Wikimedia Commons/ Jehoshaphat Aspin/*A Familiar Treatise on Astronomy*/Sidney Hall/Richard Rouse Bloxam/ Adam Cuerden)

The series of 25 cigarette cards titled *Those Pearls of Heaven* was produced in 1914 by John Player & Sons. The first five cards of the set are shown here, and are typical of the overall set which depicts both traditional constellations as well as general illustrations relating to the stars and planets. (Imperial Tobacco Ltd)

Issued by the English cigarette company W. A. & A. C. Churchman – a subsidiary of John Player & Sons – the set of 40 cards called *Howlers* was illustrated with a series of delightful cartoons created by the talented Dublin-born artist René Bull (1872–1942). The set of cards derived its name for the rather feeble jokes printed on each one. Typical of this is the card 'Astronomy', number 17 in the set. The text on the reverse includes the phrase: 'The meridian of Greenwich is a line that isn't there, kept at Greenwich to measure the time with.' (Imperial Tobacco Ltd)

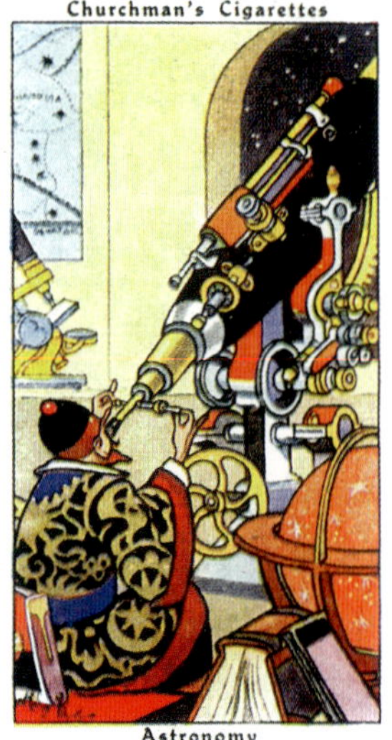

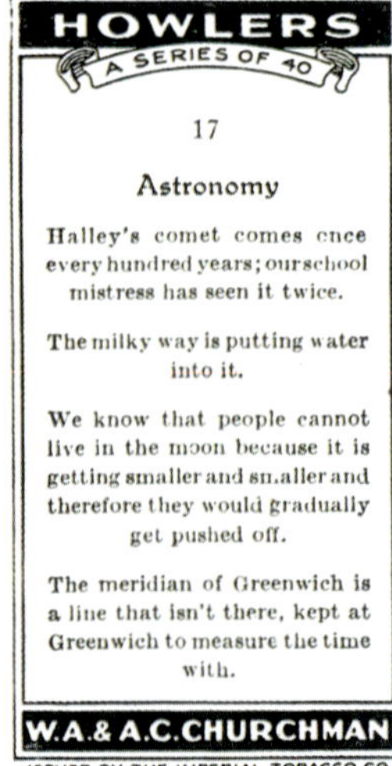

The stunning *Visions of the Universe* set of stamps was issued by Royal Mail in 2020 to celebrate the 200th anniversary of the Royal Astronomical Society. Illustrator Robert Ball's pictures celebrate astronomical objects and phenomena associated with British astronomers' discoveries. (Stamp designs © Royal Mail Group Limited)

The *Astrophilogeon* and *Urania's Mirror* cards illustrate that there has always been a market for astronomical and astrological imagery, which also explains the appeal of various collectible card sets over the years. Tobacco companies used to include cards in cigarette packets to encourage brand-loyalty and to drive sales, including two astronomical sets featuring constellations, planets,

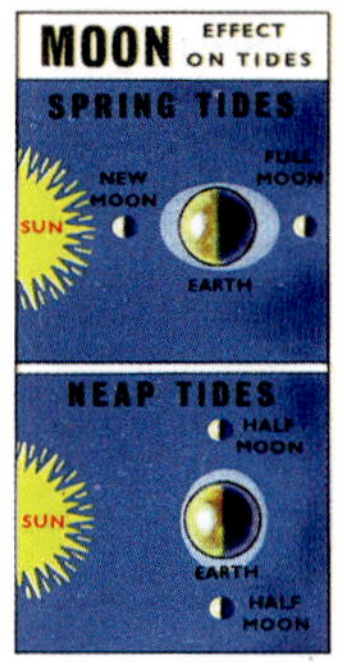

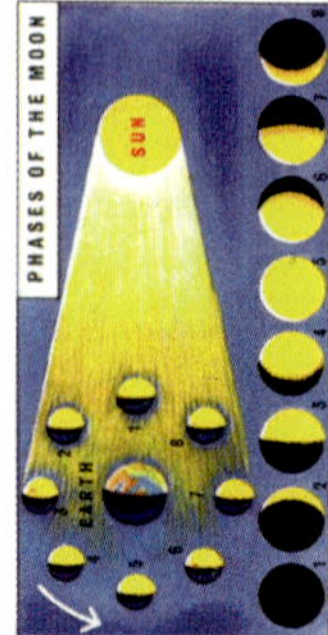

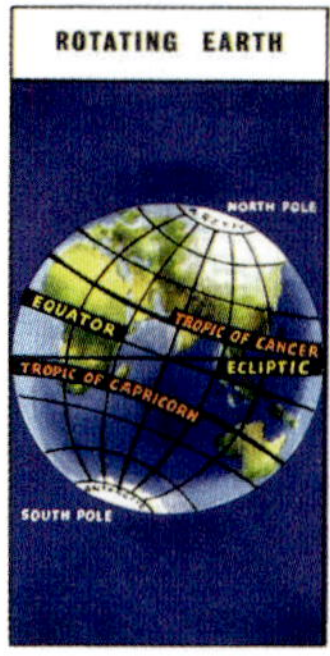

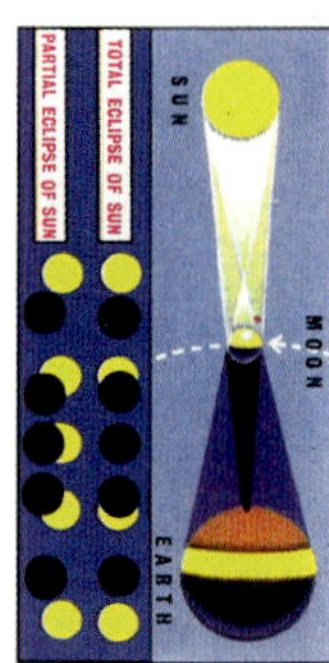

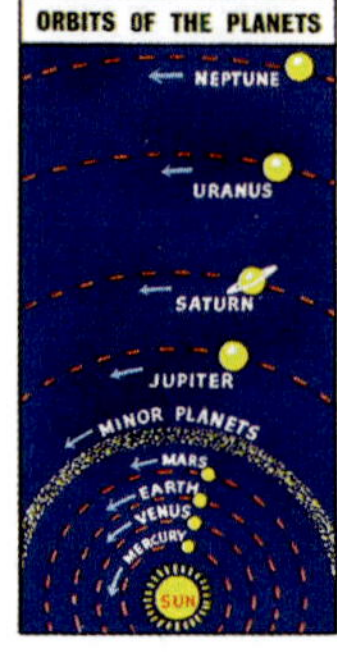

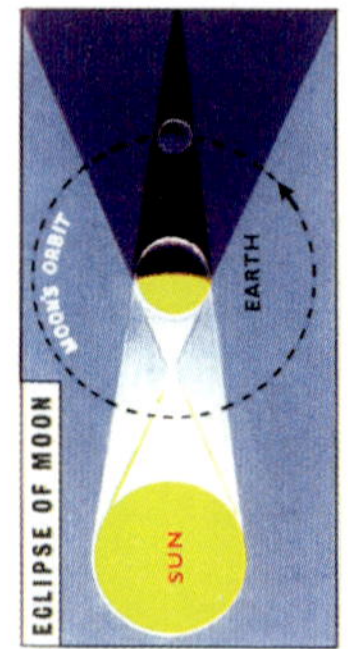

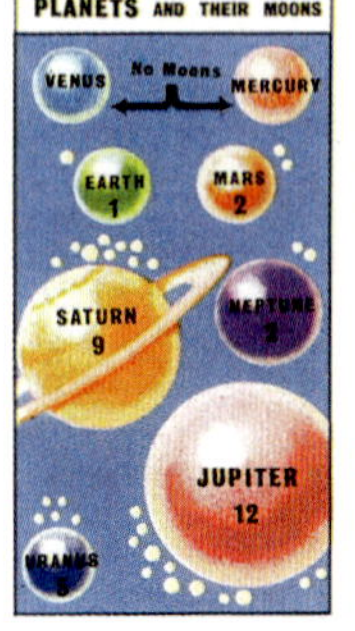

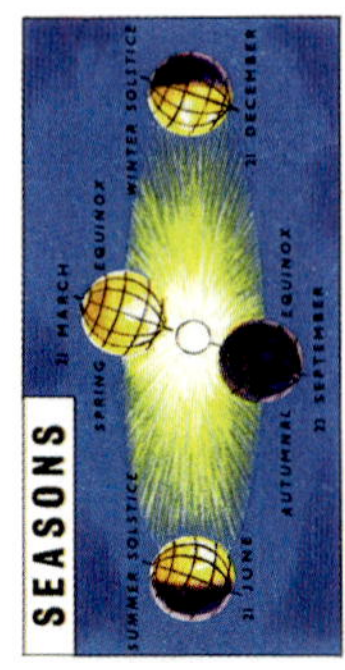

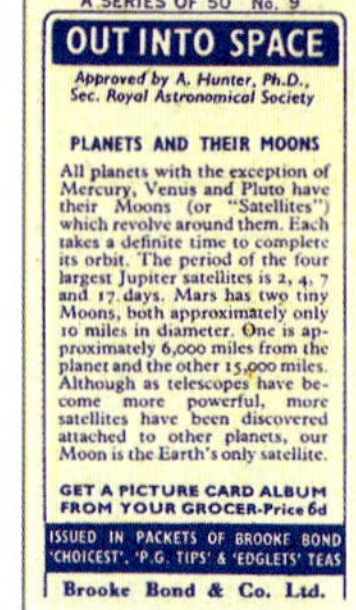

The *Out Into Space* tea cards were issued in 1956 and 1958 and given away in packets of Brooke Bond PG Tips. The set employed the usual 50-card format and, as with all other cards issued by the company, display albums were available for purchase. The cards illustrated a number of astronomical concepts including the seasons, lunar phases and eclipses, as well as the planets and their moons, comets and meteorites, stars and constellations and the Milky Way. Astronomical instruments such as radio telescopes and the astrolabe also featured. (Lipton Teas and Infusions/ ekaterra/Brooke Bond & Company)

and brief explanations of phenomena such as eclipses. In 1914, John Player & Sons cigarettes included cards from a 25-card collection called *Those Pearls of Heaven*. These mixed astronomical and astrological imagery and ideas freely, and the card-backs each bore a paragraph of explanatory text.

Some ten years later, the Wills Cigarette Company used a very similar mixture of images and text for its *Romance of the Heavens* set of fifty cards. These included some stunning artist's impressions of the surface of other planets, along with diagrams of contemporary scientific theories of planetary and lunar formation.

By the 1950s, cigarette cards had fallen out of fashion. The fifty card format was revived in 1958, this time as collectable sets in packets of Brooke Bond tea! The *Out into Space* collection also included a wide mix of topics, but every card proudly proclaimed that its text had been approved by Dr Alan Hunter, who at the time was Secretary of the Royal Astronomical Society and who spent his career at the Royal Greenwich Observatory, serving as Director between 1973 and 1975.

Of course, such trading cards are not the only small collectable; coins and postage stamps also provide a rich seam of astronomical imagery (although they are more likely to honour people, including astronomers and astronauts, which is a little beyond the scope of this chapter). Stars, associated with the heavens and hence divinity, have appeared on coins since at least Roman times, as have comets are symbols of change. A particularly bright comet was observed in 44 BCE, shortly after the assassination of Julius Caesar, and it was quickly adopted into the symbolism of Roman coinage to represent the divinity of the newly-Imperial family.

Ian Ridpath, himself an astronomer and author, has investigated postage stamps extensively, finding constellation designs on Brazilian stamps dating back to the 1880s (relatively soon after the concept became adopted internationally). British designs in the twentieth century have ranged from a picture of Jodrell Bank's iconic radio telescope, to special issues marking particular events such as the appearance of Halley's Comet, or eclipses visible from the UK. More recent stamps have used pictures from the Hubble Space telescope and assorted European Space Agency missions.

Imagery of the heavens, and the words and names of objects and concepts from astronomy, surrounds us in our everyday lives, in subtle or more obvious ways. Humans have been fascinated by the appearance of the night sky from prehistoric times, and this has been reflected throughout ancient and modern times in varied and sometimes unexpected contexts. It will doubtless continue to inspire designers, advertisers, business leaders, and even confectionary producers for many years to come.

Astronomy in the Mass Media

David M. Harland

Astronomy, like dinosaurs, is a topic that appeals to the public, in part, perhaps, because it does not have a 'downside', it simply captures the imagination.

It benefits from professionals serving as popularisers, for example, in their unique ways at different times, Carl Sagan in America and Fred Hoyle in Britain.

There is an O-Level for astronomy, but the subject is better suited to universities, where it is often incorporated with physics as astrophysics. And to become a professional you will likely need a doctorate, so it will take the better part of a decade at university simply to reach the point of qualifying to apply for the funding required to pursue research. Just as for particle physics, the

Placing observatories atop mountains away from the 'light pollution' of cities gave an unrivalled view the Milky Way, a galaxy in which the Sun is merely one of several hundred billion stars.. In this case the dome hosts the 3.6 metre telescope of the European Southern Observatory at La Silla on the outskirts of the Atacama Desert of Chile at an altitude of 2,400 metres. (ESO/S. Brunier)

facilities for astronomy are expensive and are often pursued as international collaborations.

On the other hand, astronomy is one topic where amateurs are able to work in tandem with professionals, using their abundance of time and familiarity with the sky to alert major observatories to interesting events such as the appearance of comets and novas. In fact, there was a time when amateur astronomers played a leading role.

The Royal Greenwich Observatory was established by Charles II in 1675 with John Flamsteed, Astronomer Royal in residence, charged with compiling an accurate map of the sky to assist with navigating the oceans, and mapping remained its *raison d'être* into the Victorian era. By that time, however, the innovative astronomical work was being done by gentleman of private means who placed powerful telescopes in their gardens, or even in their houses, and in some cases equipped them with spectroscopes to undertake pioneering studies.

After it was discovered that the 'seeing' was better at high altitude, observatories were built on the summits of mountains, and that most precious of commodities, observing time, was competitively granted to astronomers from institutions around the world who, upon arrival, had to acclimatise to the thin air. Nowadays, however, many astronomers work remotely using the internet with the telescopes being operated by specialist employees of the observatories.

Much astronomical research is highly specialised but there are some big questions, and the public is particularly fascinated by the possibility of extraterrestrial life, so let's consider two examples, one involving the Moon and the other Mars.

It was on 15 January 1834 that the English astronomer John Herschel arrived in South Africa, where he set up a 21-foot focal-length reflecting telescope near Cape Town on a mission to extend into the southern skies the celestial survey carried out by his esteemed father William.

On 21 August 1835 the New York newspaper *The Sun*, in its second year of publication, announced its plan to reprint

A portrait of astronomer John Herschel following his return from South Africa, during which he extended into the southern skies the celestial survey carried out by his father William. During his time in South Africa, a New York newspaper ran a hoax story that he had discovered strange creatures living on the Moon. (Wikimedia Commons/Henry William Pickersgill)

An illustration printed by the New York newspaper *The Sun* in 1835 purporting to show creatures on the Moon, including bat-like winged humanoids, which it claimed had been seen by the astronomer John Herschel in South Africa using a telescope of unprecedented power. The story attracted considerable public interest, but when it was revealed to have been made up it became known as 'The Great Moon Hoax'. (Wikimedia Commons/*The Sun*, New York City)

a series of six articles which, it said, were based on a *Supplement to the Edinburgh Journal of Science* reporting discoveries made by Herschel in Cape Town with "an immense telescope of an entirely new principle".

The series began on 25 August and caused a sensation by describing, in all seriousness, oceans, beaches and vegetation on the Moon, and the presence of animals, some familiar such as bison and goats and others mythical such as unicorns, bipedal tail-less beavers and bat-like winged humanoids who were clearly civilised.

Supposedly these accounts were based on a narrative provided by Herschel's travelling companion and secretary Dr. Andrew Grant.

Once its circulation had been greatly boosted, *The Sun* reported the sad news that the powerful lens of the telescope had inadvertently focused sunlight and started a fire that destroyed the facility.

Of course, Herschel had not invented a telescope of such power, did not have a companion named Dr. Grant, and had not discovered life on the Moon.

Several years later the newspaper's reporter, Richard Adams Locke, admitted to having conceived and carried out what had become known as the 'Great

Moon Hoax'. Locke may have been inspired by a report a decade earlier by Franz von Paula Gruithuisen at Munich University of features on the Moon that suggested to him the presence of vegetation and intelligent beings. Locke merely used the fact that Herschel, one of the most famous astronomers of

The War of the Worlds was serialised in *Pearson's Magazine* in 1897 prior to being issued as a novel by Heinemann the following year, but H.G. Wells preferred the artwork by Henrique Alvim Corréa for the 1906 Belgian translation (an example of which is seen here in a scene depicting a Martian fighting machine in the Thames Valley, near London, England). The monstrosity of the creatures sowed a public fear of invading aliens that would pass down the generations. (Wikimedia Commons/Henrique Alvim Corréa)

that time, was indeed in South Africa and, given the slow communications, would not be readily contactable to contradict the fantasy being perpetrated in his name.

On hearing of the story Herschel was amused, but he grew weary of being approached by people who believed the astonishing tale.

One might despair at the gullibility of the public of that time, but fast forward to the New York of 1938 and a radio drama transmitted by the Columbia Broadcasting System. In this the focus was the planet whose bright red appearance in the sky reminded the Greeks and the Romans of their bloodthirsty gods of war: Ares and Mars, respectively.

To mark Halloween, Howard Koch and John Houseman adapted the novel *The War of the Worlds* published in 1897 by H. G. Wells expressing in fictional form a critique of the imperialism of the British Empire with London being invaded by a ruthless enemy in a reversal in the roles of oppressor and oppressed. To suit their American audience, Koch and Houseman relocated the action to New Jersey.

To mark Halloween in 1938 CBS in New York broadcast a radio play based on H. G. Wells' novel *The War of the Worlds*. Here Orson Welles, who narrated the play in which aliens were reported to have landed in nearby New Jersey, explains to reporters that no one connected with the production had any idea that their show would cause panic. (Wikimedia Commons/Acme News Photos)

It was produced, directed and narrated by Orson Welles and ran for an hour without pauses for sponsorship. It began as a normal broadcast featuring music. After a few minutes, Welles interrupted to announce that astronomers were reporting bright flashes on Mars. Moments later he reported a strange object had fallen near the small rural settlement of Grover's Mill, which was a real place. There followed a series of ever more gruesome eye-witness accounts of monsters spreading death and destruction.

At that time, people had their radios on in the background in much the same way as a later generation would with its televisions. Network radio was a trusted source of news. And in October 1938 the news was downbeat with the prospect of Nazi Germany attacking its neighbours with an unprecedented aerial bombardment.

In Grover's Mill, men grabbed shotguns to defend their community from the invaders. Of course, portable transistorised radios were still a thing of the future, so as soon as people left their homes they lost touch with the play. But this didn't matter because, as happens all too easily in a panic, those involved spread their own rumours of terrible events. As a result, the panic continued long after the show had wrapped up by cheerfully wishing its listeners a happy Halloween.

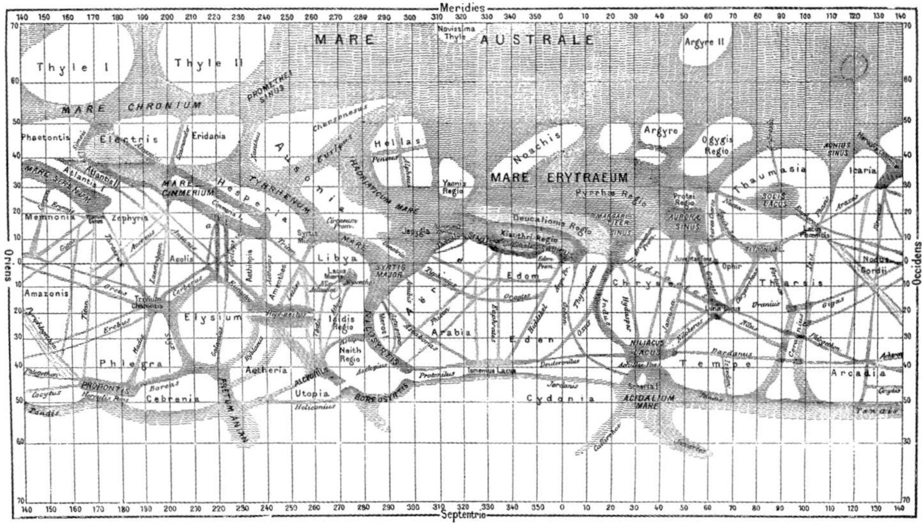

In compiling this map of Mars, based on his observations made when the planet was favourably presented during the period 1877-1886, the Italian astronomer Giovanni Schiaparelli chose a nomenclature based on classical geography (many the names are still in use today). The prolific French astronomer Camille Flammarion featured the map in his 1892 volume *La Planète Mars*, which for many years was the standard reference on the topic. (Giovanni Schiaparelli/La Planète Mars/Wikimedia Commons)

Yet again we may smile at the readiness of the public to accept news of invading alien monsters, but for a long time astronomers themselves presumed there were intelligent beings on other planets.

When the advent of the telescope revealed the planets to be worlds in their own right, the concept of "plurality" was widely believed. For example, in a scientific paper in 1784 in the *Philosophical Transactions of the Royal Society of London* discussing conditions on the surface of Mars, William Herschel referred almost casually to the natives, saying, "… the inhabitants probably enjoy conditions analogous to ours in several respects."

After studying Mars when it was particularly close to us in 1877, the Italian astronomer Giovanni Schiaparelli reported the presence of fine channels or *'canali'*. Intrigued by the English translation as 'canals', and the connotation of artificiality, the wealthy American Percival Lowell established his own observatory in Arizona and set about mapping the planet. He published a trio of books between 1895 and 1908 explaining his belief that an ancient race had built a network of canals to transport water from the polar ice caps into the arid deserts. Astronomers were sceptical but the public were captivated. At that time on Earth, the most audacious engineering project was the creation of the Panama Canal, so people could readily relate to advanced beings criss-crossing their planet with canals. It was shortly after Lowell released his first book, in fact, that H. G. Wells penned his tale of Martians seeking to escape their dying world by invading Earth.

In 1956 it was possible for Patrick Moore, who was yet to launch his remarkable career as a 'television astronomer', to publish a slim volume containing literally everything that was known about Mars. He held the prevailing view that the dark markings were vegetation undergoing an annual growth cycle.

Nevertheless, when a spacecraft returned the first close-up pictures of the surface of Mars in 1965 some people were hoping to see ancient canals. On seeing the desolate cratered terrain, the public lost interest in the planet almost overnight. But interest returned in 1997 when people saw the rover named *Sojourner* trundling around its lander, examining rocks. This mission set a record for 'hits' on the internet. In the ensuing years, a fleet of probes mapped the planet from orbit in such detail that oceanographers wryly noted that we were now more familiar with its surface than we were with the floors of our oceans.

Harking back to the Moon, when the Soviet Union was first to land a probe there in 1966 the transmission was picked up by the Jodrell Bank radio telescope and decoded, with the result that the *Daily Express* was able to 'scoop' the official release of the historic first image of the lunar surface.

The March Equinox 2012 issue of the *Planetary Report* featured a vision of Mars that appeared in the 1908 book *Mars as the Abode of Life* by the wealthy American astronomer Percival Lowell, who believed an advanced race of beings had created a network of canals to transfer water from the polar ice caps to the arid deserts. (The Planetary Society)

Undoubtedly, the highlight of public interest in space was the night that Apollo 11 landed on the Moon in 1969. Both the BBC and ITV ran extensive 'live' coverage of the landing and the historic moonwalk that followed. But apart from the unfolding drama of whether the Apollo 13 crew would make

Public interest in Mars took a nose-dive in 1965 when it was discovered to have a cratered surface, but soared again in 1997 with the exploits of the *Sojourner* rover which drove up to a succession of rocks and positioned an instrument to analyse their compositions. (NASA/ JPL-Caltech)

it back to Earth alive, public interest in the later lunar surface missions waned, almost as if they were reruns of a television show.

When the Voyager missions set off to explore the outer planets as a prelude to departing the solar system, they each carried a metal disk on which were recorded the sights and sounds of humanity. It is doubtful that these disks will ever be played by aliens, but they may well outlast the species that made them.

More recently, members of the public have been invited to submit their names for carriage aboard spacecraft and rovers. People's ashes have been launched into space, most notably on the NASA space probe Lunar Prospector that carried the ashes of planetary geologist Eugene 'Gene' M. Shoemaker to the Moon and who, but for medical disqualification, had hoped to become the first geologist to walk on the lunar surface.

Hopefully we will return to the Moon very soon, and the moonwalks will be livestreamed over the internet; and when we venture onto the surface of Mars, millions of people will likely 'ride along' by virtue of total immersion virtual reality.

In 1958 the Harvard-trained radio astronomer Frank Drake (here seen in the early-1960s) joined the National Radio Astronomy Observatory at Green Bank, West Virginia. (NRAO/AUI/NSF)

With the realisation that there are no intelligent beings elsewhere in the solar system, public expectation has turned to the stars.

In 1960, Frank Drake fitted a large radio telescope in West Virginia with a receiver which was capable of detecting an alien civilisation in a nearby stellar system transmitting signals similar to those we ourselves were 'leaking' into space. Although the result was negative, many other observatories joined this Search for ExtraTerrestrial Intelligence (SETI) and pretty soon there was a vast amount of unprocessed data. As soon as a distributed computing capability became possible in the 1990s, the SETI@home project used the internet to send parcels of data from radio telescopes to home computers, to analyse when they would otherwise be idle. The public embraced the idea. Nowadays, however, specialised computers are capable of doing such processing on-site. With listening for radio signals showing no result, we have started watching for aliens communicating using pulsed lasers.

Some people may decry looking for signals from aliens as a waste of money, and admittedly it is a long shot, but the challenge

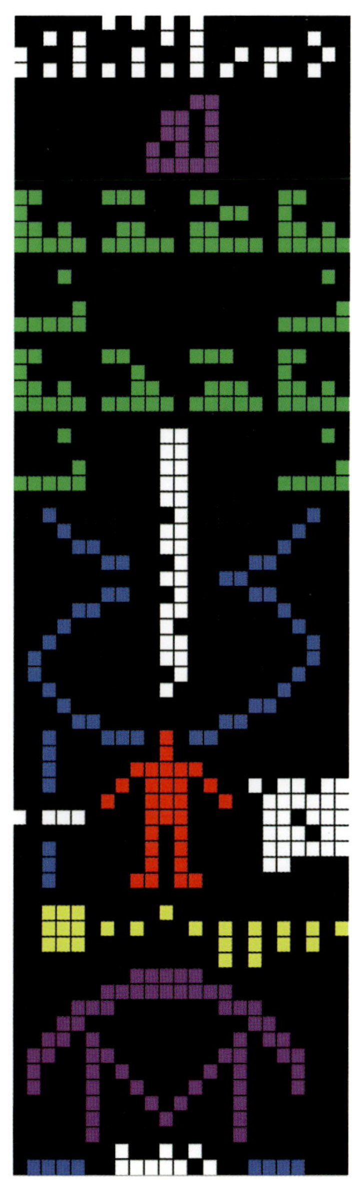

A message transmitted towards the globular cluster M13 by the 1,000-foot radio telescope of the Arecibo Observatory in Puerto Rico on 16 November 1974. The message was sent just once and lasted only three minutes. It comprised 1,679 bits of binary '0' or '1' data. This number is a product of two prime numbers and thus allows the data to be arranged as 73 rows and 23 columns which depict, from top to bottom, the numbers one to ten (white), the atomic numbers of the elements that make up our DNA (purple), the formulas for the compounds of the nucleotides in DNA (green), an estimate of the number of DNA nucleotides in the human genome (white), a graphic of the double helix of DNA (blue), the height of an average human (blue-white), the physical form of a human (red), the human population of Earth (white), a graphic of the solar system, indicating the planet that transmitted the message (yellow), and a graphic of the telescope, indicating its diameter (purple, white, and blue). It remains to be seen whether any alien receives the signal, decodes its contents and responds. (Wikipedia/Arne Nordmann (norro))

is irresistible because the payoff would be profound. As the renowned science fiction author Arthur C. Clarke once said of the search for alien intelligence, "Two possibilities exist. Either we are alone in the Universe or we are not. Both are equally terrifying."

Reversing the situation, in 1974 the Arecibo radio telescope beamed a transmission in the direction of M13, a globular cluster in the constellation of Hercules that is composed of several hundred thousand stars which, despite being some 25,000 light-years away, is one of the nearest. The transmission was the highlight of a ceremony to mark the recommissioning of the telescope after work to improve its dish, and was more of a demonstration of our ability to transmit over long distances than a serious attempt to enter into a conversation with possible extraterrestrials. The signal encoded a simple line drawing, the rationale being simply to give any aliens living there an opportunity to detect a signal that was artificial. Over the years others have made transmissions carrying more information about who we are, and where to find us using pulsars as navigational references. It has been argued, however, that we ought not to announce our presence. As the English cosmologist and theoretical physicist Stephen Hawking warned, "If aliens visit us, the outcome would be much as when Columbus landed in America, which didn't turn out well for the Native Americans."

Apart from listening for aliens, one of the most fascinating of recent astronomical developments to capture the imagination of the public is the existence of extrasolar planets orbiting in the 'Goldilocks zones' of stars.

There are compelling reasons to believe that life arises naturally in a soup of chemicals and that, therefore, potentially habitable worlds may well host forms of life equivalent to bacteria. To establish that this has happened at least once, whether it be elsewhere in our solar system or in another star system would be significant because, as Philip Morrison of the Massachusetts Institute of Technology said many years ago, this would, "… transform life from the status of a miracle to that of a statistic."

The public can keep up to date with developments in astronomy and space exploration in a variety of ways.

Astronomy has long been well served by high-quality magazines. In America there is *Sky & Telescope* and *Astronomy*, and in the UK there is *Astronomy Now* and the BBC *Sky at Night* magazine, the latter being an offshoot of the longest running programme in the world, on any topic. In addition, there are astronomy clubs all across the country and the British Astronomical Association assists amateurs in activities such as solar observing, lunar observing and variable star observing. There is also the *Yearbook of Astronomy*, with previews of things to look out for in the coming year, summaries of recent discoveries, and articles on the development of astronomy.

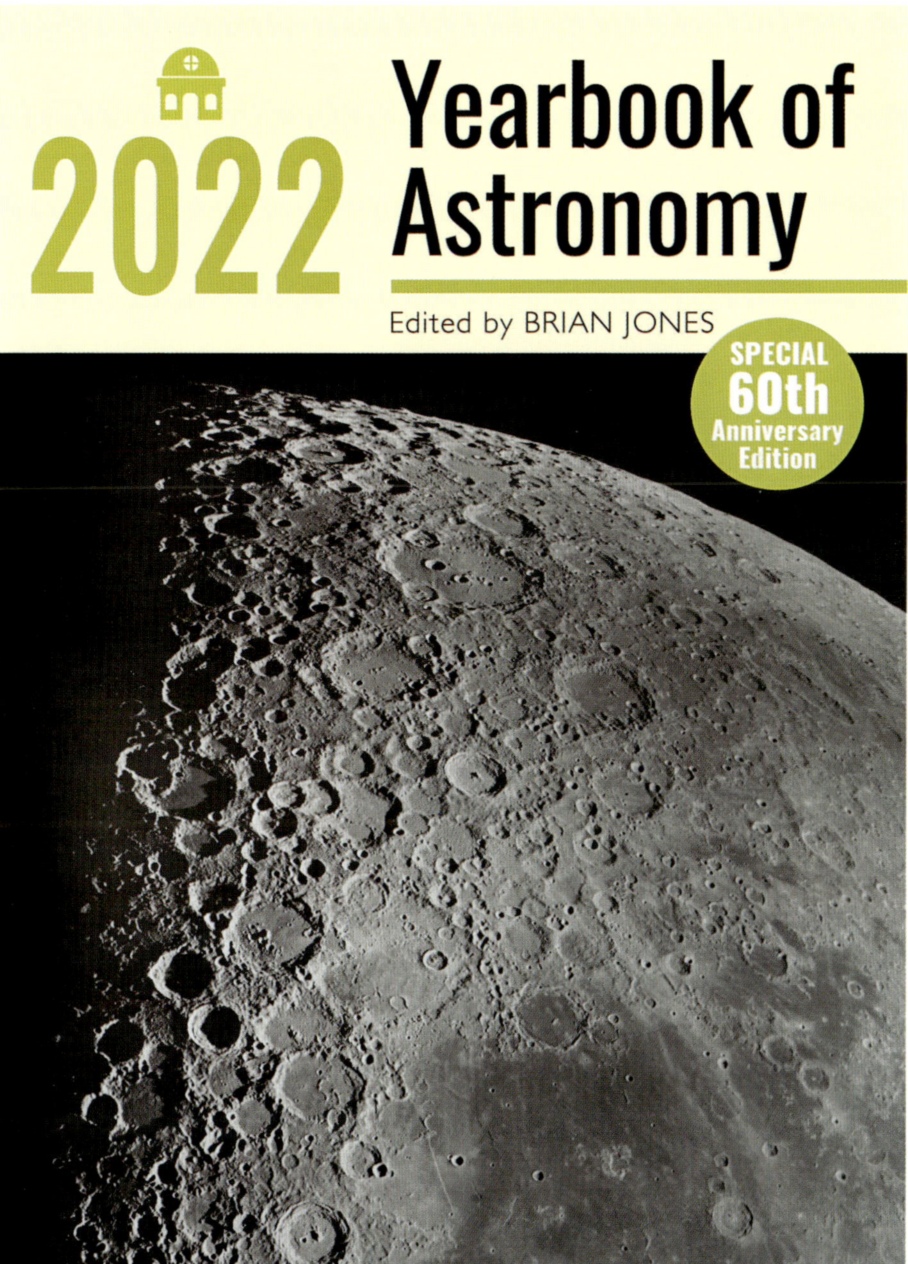

To mark the sixtieth anniversary of the *Yearbook of Astronomy* – believed to be the longest-running 'off-the-shelf' astronomical publication of its kind in the world – the 2022 edition reproduced the cover of the inaugural (1962) edition, featuring a view of the cratered highlands of the Moon that was originally taken in 1919 through the 100-inch telescope of the Mount Wilson Observatory in California. (Pen & Sword Books Limited/Yearbook of Astronomy)

For many years, the American astronomer and science communicator Carl Sagan led the way in promoting astronomy and space exploration with numerous books, public talks, and in particular his *Cosmos* television series. Nowadays, the baton has been taken up by Bill Nye 'The Science Guy' and Neil deGrasse Tyson.

As regards the space side of astronomy, there is the quaintly named British Interplanetary Society, with its *Spaceflight* magazine. In America, the Planetary Society publishes its *Planetary Report* and pursues political advocacy for projects in support of astronomy, planetary science, and space exploration.

Scientific establishments issue press releases to announce discoveries and scientists write papers for peer-reviewed journals. In the UK there is the Royal Astronomical Society, with its *Monthly Notices*. In the USA the American Astronomical Society publishes premier journals such as *The Astrophysical Journal* and *The Astronomical Journal*.

Embargoed press releases allow journalists time to use their personal contacts in the community to add to imminent announcements of the latest discoveries, seeking 'spin' to support or criticise a particular piece of work.

In addition to signing up to receive email press releases from NASA, ESA, individual scientific institutions, and societies, there are a number of astronomy outreach products. The International Astronomical Union sponsors *Communicating Astronomy with the Public*. In America, the Lunar and Planetary Institute publishes the *Lunar and Planetary Information Bulletin*. The latest results are also announced at conferences, in particular the annual Lunar and Planetary Science Conference (LPSC) whose inaugural gathering in 1970 featured the analyses of the Apollo 11 lunar samples.

The NASA Planetary Data System archives scientific data from space missions for study by professionals, and processed imagery is publicly available at the online NASA Photojournal hosted by JPL.

Unfortunately, the internet not only enables professional scientists to disseminate information to the public, it has also opened the floodgates for a lot of nonsense about astronomy and space exploration promoting conspiracy theories that NASA has secret evidence of alien artefacts on the Moon, on Mars, and basically everywhere that we have sent probes. Indeed, in an age in which the traditional concept of 'objective truth' is under threat, the internet is rife with 'proofs' of how the Apollo lunar landings were faked.

Nevertheless, the public has a fair impression of the solar system that we inhabit, and thanks in particular to the Hubble Space Telescope this awareness extends right across the universe.

Compare this to 1910, when Earth passed through the tail of Halley's Comet and the discovery that it contained a cyanide compound prompted fears that we'd all be poisoned.

The periodical journal *Communicating Astronomy with the Public* is an astronomy outreach product jointly sponsored by the International Astronomical Union (IAU) and the National Astronomical Observatory of Japan (NAOJ) in collaboration with the European Southern Observatory (ESO). It features articles aimed at astronomy communicators and bridges the scientific astronomical community with society. The background image on the cover seen here shows the Atacama Large Millimeter/submillimeter Array (ALMA) in Chile, with the centre of the Milky Way in the sky above. Back issues of the journal since publication started in 2007 are downloadable online at **capjournal.org** (IAU/Communicating Astronomy with the Public/ ESO/B. Tafreshi **twanight.org**)

After astronomers detected a cyanide compound in the tail of Halley's Comet, using spectroscopy as it approached Earth in 1910, the famous French astronomer Camille Flammarion warned that "Cyanogen is a very deadly poison, a grain of its potassium salt touched to the tongue being sufficient to cause instant death." The resulting panic was made fun of by postcards identified as *Carte officielle de Souvenir de la Fin du Monde le 19 Mai 1910* that were produced by F. M. of Cologne. From the collection of Richard Jakiel. (F. M. Cologne/Richard Jakiel)

In July 1994 the world looked on in fascination as a string of fragments from Comet Shoemaker-Levy 9 – originally discovered on 24 March 1993 by astronomers Carolyn and Eugene M. Shoemaker and David Levy – smashed into Jupiter. This celestial spectacle raised public awareness of the threat posed by the prospect of a future impact on the scale of the one that wiped out the dinosaurs. (R. Evans, J. Trauger, H. Hammel, Hubble Space Telescope Comet Science Team and NASA)

With the recent realisation that the dinosaurs were wiped out by the environmental aftermath of a major asteroid strike, astronomers have been scanning the skies for anything that might strike Earth in the near future. Whenever they find an object that will pass very close by, this attracts press attention. Unfortunately, thanks to 'disaster movies' such as *Armageddon*, the public expectation is that we will be able to launch a mission to prevent a looming 'extinction level event', whereas in reality we will merely be able to record it unfolding on our cell phones!

On an entirely different level, chocolate manufactures have long held a fondness for astronomy, with products such as Mars, Milky Way and Galaxy, as well as specialist boxes where the individual confectioneries celebrate the planets.

For a long time, astronomy and astrology were linked in the public mind. Johannes Kepler, whose early seventeenth-century discovery of the 'laws of planetary motion' put astronomy on a firm mathematical footing, had a lucrative side-line of telling fortunes although it is to be hoped he did it with

A depiction by space artist Don Davis of a large asteroid colliding with Earth. An impact on such a scale, along with the resulting extinction of the majority of the species on the planet, is classified as an 'extinction level event'. The demise of the dinosaurs, which occurred around 65 million years ago, is attributed to the environmental effects of the impact of a body a mere 10 kilometres in diameter. (NASA/Don Davis)

his tongue firmly in his cheek. Although some newspapers and magazines still publish horoscopes to allow their readers to check what is 'written in the stars', thankfully most people are aware that this has nothing to do with astronomy.

Astronomical events that gained particular press attention during recent decades include a comet colliding with Jupiter; the first-ever flyby of Pluto; the combining of data from radio telescopes to produce images of supermassive black holes at the centres of galaxies, including our galaxy; after many years of trying, seeing gravitational waves literally shaking the universe; the first pictures from the James Webb Space Telescope; and the CERN Large Hadron Collider probing the earliest moments of the universe.

As regards our probing of the Act of Creation, the Belgian priest, mathematician and theoretical physicist Georges Lemaître, who predicted that the universe emerged from what he called the 'cosmic egg', saw no conflict between science and religion, viewing "… science and faith as two different ways to know truth."

So to sum up, the public has long held a fascination with aspects of astronomy, in part for its sheer scope, spanning the entire history of the universe, and thanks to the internet it has never been easier to keep up to date with developments.

Music and the Cosmos

A Symbiotic Relationship

Jonathan Powell

The vast array of drawings, pictures, and photographic images that have been accumulated throughout history depict a dynamic, evolving and at times dramatic catalogue of events in the night sky. From vivid and colourful artwork painted inside a cave dwelling, to the extreme cutting-edge of digital imagery, stars, planets, comets, and possibly even supernovae have all been captured.

This astonishing databank of images has provided an invaluable timeline of events and a priceless wealth of information. The changing aspect of how these pictures have been obtained also charts the progress made by humankind as to how they are collected, and the corresponding advancements made in technology. In turn, as the centuries have passed, many of these images have become interwoven in our consciousness as our species pushes back the boundaries that once held us captive, reaching further and deeper into the realms of space, and as we do so, capturing that very moment.

However, far beyond the reach and influence of these fantastic and lavish pictorial representations of space, a deeper, greater, and more fundamental bond exists between us and the surroundings in which we live. It weaves, threads, and touches a plane that is not only equal to but in many regards' superior to the picture, that of music. As one marvels at the colours and sinuous texture of a vast nebula than spans millions of miles, there is also the presence with such imagery of a background musical chorus, one which has been in evidence since the birth of universe. That music resonates with its own pulse of life.

The definition of the music can take many forms and like the response of a dog to a whistle whose frequency is only tuned for the target recipient, there are a great number of levels to explore in the relationship between humankind and our environment, some perhaps acutely evident, others not so. The 'fine tuning' to discover such a vast array of different cosmic music is not dissimilar to casting more sensitive and powerful optics on objects that have already been discovered, enhancing their imagery to yield greater clarity and definition, revealing a lot more than previously revealed.

Early Understanding of Our Musical Relationship

The notion of such a musical relationship has its roots firmly entrenched in Greek culture. Any would-be scholar in ancient Greece would find that alongside arithmetic, geometry, and astronomy, (referred to as 'spherics'), music had a place of equal rank. The four disciplines, the *quadrivium* as it was referred to, would form the basis of the curriculum that every pupil would need to master before advancing to pursue their chosen specialised subject.

Pythagoras and his followers who believed that 'Number Rules the Universe'. (Wikimedia Commons/ Musei Capitolini, Rome/No. 53/ Photo by Szilas, 2013-03-04)

More than 2,500 years ago, it was the Greek philosopher Pythagoras, (c. 570–c. 495 BCE), who applied his own interpretation in music theory to the behaviour of the celestial objects. Whilst experimenting with taut strings, Pythagoras discovered that shortening the length of a string to one-half of its original length, raised the pitch at a harmonious interval, an octave.

When plucking the string of a musical instrument, we know that the pitch of the note produced is related to the length of the string that is being played. Introduction of a second string and ratios between string lengths and notes achieved are thus formed, for example, 1-2, 2-3, 3-4. Pythagoras believed that such a harmonious musical balance as achieved by the use of chords in such a manner existed in the universe, and whilst inaudible to human ears, he characterised the intervals between the orbits of the planets of the solar system as either tones or half-tones, with the addition of up to seven whole tones that in-turn created a synchronised natural order. Pythagoras also believed that the Sun, Moon, and planets, all emit their own unique 'hum' based upon their orbital revolution.

Just as certain mathematical patterns underlie the musical scales and intervals that perform a melody that is most pleasing to the ears of its recipient, they also, on a much deeper and more profound level, describe the probability waves that rest at the very core of quantum theory. Pythagoreans, having embraced the notion, concluded that everything in the Universe was governed by simple ratios of whole numbers, creating the Pythagorean motto, 'Number rules the Universe'.

The music discipline, part of the *quadrivium*, had *musica universalis*, 'music of the spheres' as an integral part of its teachings. It was this philosophical concept that underpinned the beliefs of the Pythagoreans, with its influence reaching far beyond those who studied it, touching on many other cultures and schools of thought during that time period.

One of the last Pythagoreans was German astronomer Johannes Kepler (1571–1630). Kepler was a scholar keen to keep alive the concept of *musica universalis* and with an eye for writing music, it is no surprise that he was to spend half of his life working on the laws of planetary motion and the possible association of musical harmony.

Johannes Kepler: A Pythagorean whose belief in the music harmonies of the universe underpinned his work on the laws of planetary motion. (John McCue)

In one of his many publications, Kepler attempted to explain the proportions of the natural world in terms of music, and in particular those in the subjects of astrology and astronomy. At the time, there were no scientific and ultimately defining lines between astrology and astronomy, with Kepler noted for studying both. The publication in question, *Harmonices Mundi*, 'The Harmony of the World' (1619), had at its core of 'harmonies' that of *musica universalis*. The final section of the book centres on Kepler's work toward his discovery of the "third law of planetary motion"

Kepler believed that each of the planets in our solar system and its respective motion around the Sun played a certain melody. Disregarding the fact that sound cannot travel through the vacuum of space because there are no particles to 'carry' said sounds, Kepler believed that human ears were unable to discern such music anyway, as the range of the notes were below audible frequencies. Kepler, in his attention to detail, even assigned each planet its own celestial tune.

Listening to the Cosmos

For centuries, observational astronomy and the great array of imagery commanded and still commands a front row seat in the dynamics of the

Giant Dish Antennas of the Karl Jansky Very Large Array (VLA). Named in honour of Karl Guthe Jansky, the VLA in Socorro, New Mexico, continues to listen to the heavens. (NRAO/AUI/NSF/J. Hellermann)

subject, but thanks to significant advances in radio astronomy, the balance is being slowly being redressed so that the lion's share governed for generations by optical astronomy now finds its dominance suitably tempered.

Radio astronomy and optical astronomy both examine the electromagnetic radiation originating from beyond Earth's atmosphere. The difference in the collection of data is the method and tools used to detect this radiation and, in the wavelength, or frequency of the waves studied. Light and radio waves are both manifestations of the same energetic phenomena but because radio waves are much longer in nature than optical waves, the telescopes used to detect them must be significantly larger than their optical counterparts.

The birth of radio astronomy dates back to the early 1930s, when American physicist and radio engineer Karl Guthe Jansky (1905–1950) first discovered radio waves emanating from the Milky Way. Jansky's curiosity to determine the location of the source saw him construct a steerable antenna and begin searching the heavens. His search method was to take directional measurements in order to pinpoint the location.

The waves had been registered on receivers operating in the 20 MHz region of the radio spectrum, (about 15-metre wavelength), and to Jansky's surprise his investigations revealed that the 'noise' was extra-terrestrial. Upon making this discovery, Jansky approached the wider science community but they did not recognize his findings, in fact becoming very dismissive of what he had actually achieved. However, later, upon reading Jansky's findings and picking up the thread of his work, Grote Reber (1911–2002), a keen radio enthusiast, speculated that the signals Jansky was hearing were in fact of thermal origin, caused by very hot objects, and as such should be easier to detect at higher frequencies. Therefore, Reber set out to narrow the effective beam width.

In the summer of 1937, Reber was ready to unveil his significantly improved version of Jansky's apparatus, revealing the first parabolic reflector radio telescope. By September that year, following a great deal of work all of which had been undertaken by Reber at his own expense, the telescope was finally ready and with its completion, the

Radio telescope pioneer Grote Reber whose work revealed radio sources in supernova remnant Cassiopeia A. (John McCue)

world had its first radio astronomer. Built in his backyard, the telescope was an astonishing achievement and for a decade or so, Reber was to rule the radio waves as the only astronomer of his kind.

Despite his tremendous effort and dedication, one of Reber's first attempts with a newly constructed receiver failed to detect signals from outer space. Undeterred, he built another receiver which also failed to yield any results, but his persistence paid off with a receiver operating at 160 MHz proving successful in 1938, confirming what Jansky had discovered.

In 1941, Reber went on to detail the first radio map of the galactic plane with further observations completed over the next two years, thus extending the detail and making the map more comprehensive. As well as his published work which included 'Cosmic Static', Reber's data was published as contour maps. These maps showed the brightness of the sky in radio wavelengths, also revealing the existence of radio sources in Cygnus A and Cassiopeia A for the first time.

Aside these radio galaxies as they became known, some cosmic objects produce signals that can be considered as a point source, for example, pulsars and quasars. Many other radio sources exist, each of which carry a signature unique to its sender. The task of the radio telescope is to receive all the sound

Situated in the Perseus Arm of the Milky Way is the supernova remnant Cassiopeia A. Captured by NASA's *James Webb Space Telescope (JWST)*, the image dramatically reflects the leftovers of a star's death. The event, which would have created a new star-like object in the heavens, may well relate to a sixth magnitude star catalogued by the first Astronomer Royal, John Flamsteed, in 1680. Whilst other notable astronomers including Caroline Herschel may also have inadvertently noted the position of this 'star', the remnant has remained the focal point of great interest as the strongest source of radio emission in the sky. (NASA, ESA, CSA, STScI, D. Milisavljevic (Purdue University), T. Temim (Princeton University), I. De Looze (University of Gent))

and noise from the object and near vicinity, then, via fine tuning mask out the sounds which do not relate to that specific object, in order to get a true sense of what is being heard. Only from this individual sound alone, can any determination be made as to the radio properties and sole identity of the object.

In our solar system, Jupiter is a powerful source of natural radio waves that can produce exotic sounds when picked up using simple antennas and shortwave radios. The sounds generated are ones naturally produced by plasma instabilities in Jupiter's magnetosphere. These plasma waves give rise to the laser radio signals that travel away from Jupiter's magnetic poles in cone-shaped beams. These beams rotate with Jupiter every 9 hours and 55 minutes making Jupiter not unlike that of a slow-turning pulsar. When the beams sweep past Earth, listeners can pick up the Jovian radio bursts in the shortwave bands between 15 and 40 MHz.

Various space probes dispatched to parts of the solar system have also heard some quite bizarre sounds. NASA's *Juno* spacecraft captured the apparent 'roar' of Jupiter, with the bow shock of crossing Jupiter's immense magnetic field recorded by the craft during a two-hour period in June 2016.

Plasma waves, like the roaring surf of the ocean, were recorded by NASA's Electric and Magnetic Field Instrument Suite and Integrated Science (EMFISIS) instrument onboard the *Van Allen Probes* which were launched in 2012. The probes also picked up strange whistles as the result of plasma waves interacting with the Earth's own magnetic fields.

NASA's *Cassini* spacecraft recorded eerie sounds from Saturn, generated by auroras near the poles of the planet. These aurorae are similar to Earth's northern and southern hemisphere lights. *Voyager* also noted evidence of lightning deep within Saturn with radio waves yielding a crackle discharge. Static was recorded by *Cassini's* Radio and Plasma Wave Science instrument as it crossed the plane of Saturn's rings in December 2016.

Many Sections of the Cosmological 'Orchestra'

As with an orchestra and the different sections that contribute to an overall sound, or indeed, as part of a solo piece, there a number of contributors that offer renditions to be reckoned with. As with the layout of an orchestra to generate the very best sound for its intended recipients, the Cosmos too promotes its own majestic sweep of playing positions that universally reach perhaps more than just one set of recipients.

Our Sun is not a just an enormous ball or light and heat; it too is a source of sound. Thanks to data collected over several decades by the European Space Agency's (ESA), and NASA's *Solar Heliospheric Observatory* (SOHO), researchers at Stanford Experimental Physics Laboratory were able to make it possible for us to hear our nearest star, and oddly enough, the provider of light and heat produces a low, pulsing 'heartbeat' type of sound. Analysing the Sun's natural vibrations and filtering out noise generated by SOHO's movements,

The Van Allen Belts form a zone of energized charged particles that surround the Earth. Named after James Van Allen, the American physicist who discovered them in 1958, the two belts comprise regions of trapped radiation that protect the Earth from harmful solar and cosmic rays. Specially designed antennas onboard NASA Van Allen Probes picked up a high-pitched chorus of chirps and beeps emanating from the belts. Whilst protecting the planet, the belts also pose a significant risk to astronauts passing through them. Intense solar activity can not only generate displays of the aurora borealis when they interact with the belts, but also diminish the outer region of the second belt. (NASA/Van Allen Probes/Goddard Space Flight Center)

researchers sped the data up by a factor of 42,000 to bring sound up to audible human levels.

The aforementioned Van Allen Belts and the behaviour of the high energy electrons produce what NASA dubbed as the 'whistler-mode chorus' as they sound not unlike something one might hear if playing a 1980s-style arcade game. Chiefly comprising an inner and outer belt, whilst the inner one is relatively stable, the outer tends to swell and shrink over time.

From Earth's protective shield to a force that cannot be escaped, the gaping maul of a black hole. At the centre of the Perseus cluster of galaxies, a super-massive black hole some 250 million light years away shouts a cosmic wail courtesy of acoustic waves propagating through the copious amounts of gas surrounding the phenomenon. In order that the sound be audible to humans, the acoustic waves have been transposed up 57 and 58 octaves. Ironically (for a black hole) the result is not only somewhat 'spooky' but a little 'angry' in nature, almost like a warning buoy placed deep in space. These eerie 'chirps' certainly add to the diversity of the sounds being generated by the cosmos.

Resonation with the Cosmos on a Spiritual Level

According to the Big Bang theory, *Om*, (from Sanskrit which means 'all'), is the cosmic sound which initiated the creation of the universe. For those involved with meditation and yoga, the chanting of *Om* will seem standard practice, but to those not involved, an explanation would be a useful first step in its understanding.

As with the word *omni* (from the Latin *omnis* meaning 'all'), *Om*, when chanted, conveys the concepts of 'Omniscience', 'Omnipresence' and 'Omnipotence', and is used at the commencement of a yoga exercise to focus the mind in preparation for meditation. *Om*, a sacred *Mantra* (from Sanskrit for 'sound tool'), is a representation of us and the universe in which we live. It is believed that mantras hold within them the latent forms of the universe.

The whole 'Om' principle is based on not one but three sounds, creating the 'Aum' sound, which in turn, ('a', 'u', and 'm'), refer to past, present, and future. To be clear, mantras are syllables that exert an influence through sound vibration that resonates on specific parts of the human body.

The human hearing range is 20 Hz to 20 KHz, with Aum vibrating at 432 Hz with the belief that this frequency is the basic sound of the universe and therefore by its chanting, we are in essence fusing and intertwining with the cosmos. It is believed that before the creation of the universe, the humming of energy existed, that in turn, became the vibration of creation, and therefore the universe was created out of the ever-present sound of Aum. That cosmic sound still exists around and inside everyone.

A whole spectrum of studies has been conducted by the science community as to the potential effects of chanting the Aum mantra, but if just one person finds well-being from doing so, then perhaps it remains a very personal approach to making that special cosmological connection.

Music Written for the Cosmos - Classical

There have been many musical works written which attempt to capture the universe in which we live with striking, unique, and ultimately defining sets of chords that readily invoke a link between

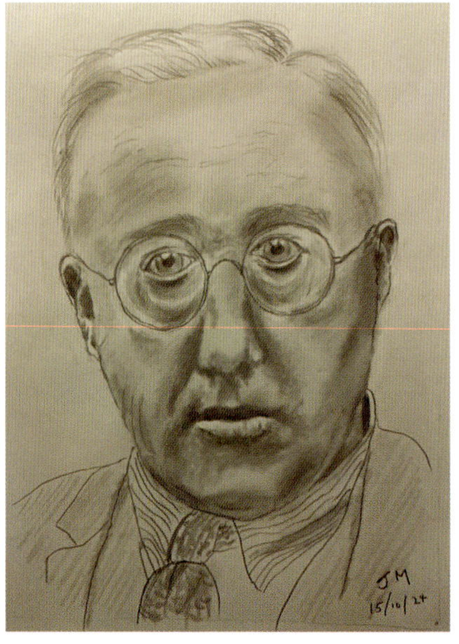

Best known for his magnificent 'The Planets' suite, composer Gustav Holst dramatically brought the solar system to musical life with a series of stirring and memorable pieces. From 1914 to 1916, starting with Mars and finishing with Mercury, his work captured the symphonic traits of each planet, making them as musically individual as their visual appearance. (John McCue)

Followed by numerous and more lavish imagery in the years that followed, this simple cover of the 1921 first edition sheet music of The Planets by Gustav Holst marks the entrance of one of the most prolific and memorable orchestral works which unlike the changing pictorial representations of its cover, remains as insightful, inspiring and unique, as the time period of which saw its creation. The seven-movement suite reflected the known planets in the solar system at the time and with its first public performance a year earlier in 1920, the brilliance of Holst's work saw him duly acknowledged as a major composer. (Wikimedia Commons/Faber Music/Gustav Holst/Goodwin and Tabb Ltd)

the individual style of a composer and their composition. No more so probably than that of English composer Gustav Holst (1874–1934), whose work *The Planets* takes us on an epic voyage of musical discovery touching on the greatly differing bodies within our solar system. Holst's idea for such a work was conceived in 1913, partly fuelled by his interest in astrology, with the actual writing of the movements commencing in 1914 with Mars, ending in 1916 with Mercury.

Each piece uniquely celebrates the individuality of each planet; its outward appearance in 'character', diverse colours, and composition. Written before the discovery of Pluto, the tribute to the planets by Holst is both remarkable in its perception and ultimately memorable in its seemingly timeless delivery. However, long before Holst, it was another great composer – Franz Joseph Haydn (1732–1809) – who penned one of the earliest operas with space in mind, and not just space, but space travel. *Il Mondo della Luna*, written

An oil on canvas painting of William Herschel, created in 1785 by the English portrait painter Lemuel 'Francis' Abbott. Visits made by composer Franz Joseph Haydn to William Herschel's observatory had great influence on Haydn's compositions. (Pixabay)

in 1777, which translates as 'The World on the Moon' creatively envisaged humankind venturing beyond Earth's atmosphere.

Haydn's association with William and Caroline Herschel, and a visit to their observatory in 1792 may well have influenced other work, including Haydn's *The Creation*. Some Haydn scholars believe the *The Creation*, possibly gives a significant nod in the direction of the nebular hypothesis, an idea that had attracted Haydn's attention. The hypothesis suggests that the Sun and Planets of our solar system were formed from coalescing gases of a swirling nebula.

Music Written for the Cosmos – Science Fiction Transitioning to Science Fact

From great classical tributes which via simple or dramatic orchestral renditions captured the composers passion and feeling for the cosmos, the baton passed into the hands of contemporary artists who in their own unique style cast

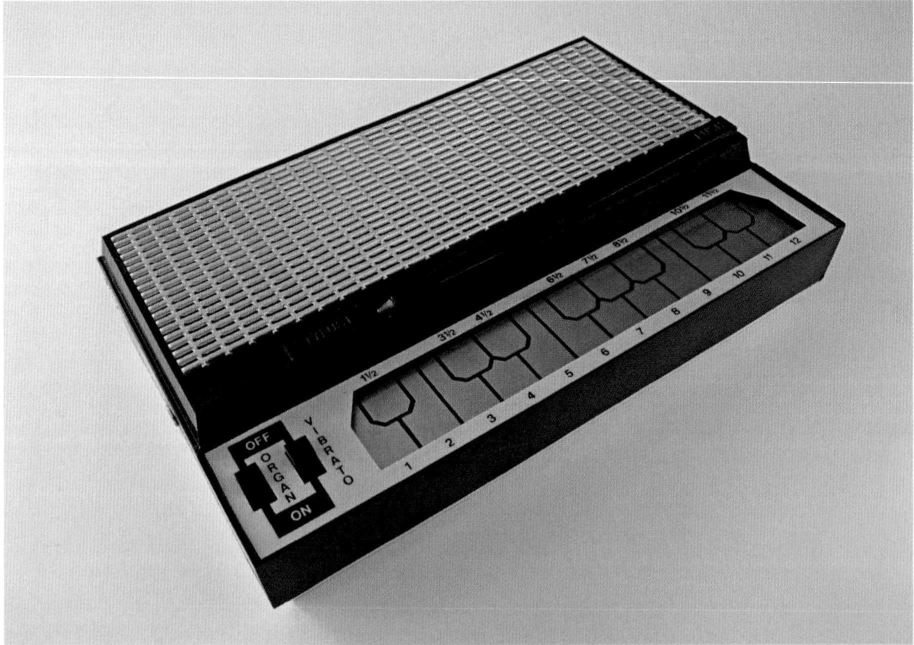

Stamping an unmistakeable presence on several music tracks particularly in the 1970s was the sound of the vintage mini-synthesizer, more widely known as the Stylophone. David Bowie, musician and popular music icon, brought the use of the instrument to the fore in the 1972 track, 'Space Oddity', which was inspired by Stanley Kubrick's science fiction masterpiece, '2001: A Space Odyssey'. Other artists of note to incorporate the instrument into their repertoire include Blondie and German electronic band Kraftwerk. The Stylophone, which chiefly saw its best sales in the children's toy market, was invented by Brian Jarvis of Dubreq Studios, London, in 1967, entering production in 1968. (Adobe Stock/robotcity)

many alternate opinions on the universe. The vast array of genres that seek to make the connection with the cosmos is as both wide as it is varied.

However, during a particular period in the 1960s where humankind placed not only craft but people in space for the first time, the era generated the dawn of concept versus reality, as boundaries were breached and science fiction became science fact. A curtain was slowly being lifted from where initially only a rather narrow shaft of light had been thrown on the capabilities of humankind. Now, a much brighter and more radiant beacon shone forth, the reality of conquering space itself.

The Canadian rock band Rush transposed the discovery of the first black hole in 1964 into music, as 'Cygnus X-1' waxed lyrical about a journey by the spaceship *Rocinante* to the hole, all of which did not end well for the craft. Cygnus X-1 comprised two parts, *Book 1: The Voyage* (1977), and *Book II: Hemispheres* (1978). Potentially seen galloping to its fate, *Rocinante* was not only the name of the horse that in Greek mythology Zeus rode, but also Nobleman Don Quixote's steed.

David Bowie (1947–2016) left a critically acclaimed legacy with his contribution to his own style of capturing the Cosmos. Embracing 'Space Oddity' (1969), 'Life on Mars?' (1971) and 'Starman (1972), the arrangements reflect the singer's vision and tuneful interpretation of so many different elements. 'Starman' was one of the tracks from the legendary album 'The Rise and Fall of Ziggy Stardust and the Spiders from Mars' and is thought to have been an influence for Steven Spielberg's 'Close Encounters of the Third Kind'.

Released in 1971, Pink Floyd's album 'The Dark Side of the Moon' features the track 'Time', which uses the image of the Sun racing to rise and set, illustrating the exorable passage of time in human life.

Music Written for the Cosmos – Contemporary

The B-52's 'There's a Moon in the Sky (Called the Moon)' (1979) asked the question as to what to call the Moon, as it has no name. It was released in the same year as the song 'Walking on the Moon' by The Police.

Sirius, the brightest star in the heavens, was the subject of the Alan Parsons Project in 1982, with Queen guitarist and astronomer Brian May producing 'Star Fleet' in 1983, for the Japanese science fiction series 'X Bomber'.

Soundgarden's 'Black Hole Sun' (1994) lyrically explored the contrast between light and dark, with songwriter Chris Cornell fascinated that a black hole can be billions of times larger than our Sun, and yet it is seen as a dark void of nothingness.

In 2002, Moby alluded in his lyrics that the human body is made up of a tiny percentage of stardust with the track 'We Are All Made of Stars'.

In 2013, Janelle Monáe's album 'The Electric Lady' features the track 'Sally Ride', a tribute to Dr Sally Kirsten Ride (1951–2012), the first American woman to reach outer space, as crew member onboard *Space Shuttle mission STS-7* in June 1983.

Released in 2021, Coldplay's album 'Music of the Spheres' was the band's second space-themed concept album, with the title embracing the Greek philosophy.

Where the Music Takes Us

During the various crewed missions into space there have been numerous reports of sounds and noises that at first naturally seemed rather baffling. The crew of Apollo 10 reported a whistling type of sound as the astronauts rounded the far side of the Moon in 1969. However, the whistling was merely the result of cross-wire communication between the lunar module and the command module. In what was a test-run for Apollo 11's actual playing of the

The crew of Apollo 11 took a selection of music on their historic trip to the Moon. (NASA)

NASA's Mars Curiosity Rover transmitted the first interplanetary music back to Earth. (NASA)

track on the lunar surface later on, Frank Sinatra's *Fly Me to the Moon* made its first trek into space with the Apollo 10 crew.

The Apollo 11 crertfw made their own musical compilations to take on their trip to the Moon. Neil Armstrong's 'mixtape', (a self-composed mix of favourite music), featured tracks from the 1947 album *Music Out of the*

Moon by Lex Baxter. Other artists to travel to the Moon in musical form only included Peggy Lee, Barbra Streisand, and Glen Campbell.

In 1988, the crew of the Soviet Soyuz TM-7 mission took Pink Floyd's live version of their *Delicate Sound of Thunder* album along with them. A claim was made by Pink Floyd band member David Gilmour that this was the first rock music recording to venture into space. However, that claim was duly disputed, with history recalling that rock music had previously been carried onboard Apollo 15. The Space Shuttle crews received 'wake-up' music, much of which was chosen by the astronauts themselves. The idea of this 'wake-up' music was continued by crews onboard the International Space Station.

Music has also been beamed back at Earth, courtesy of NASA's Mars Curiosity rover. A track entitled *Reach for the Stars*, especially penned by musician Will.i.am, was transmitted back from the surface of the red planet. Travelling over 225 million kilometres back to Earth, the recording made for the first tune to be broadcast from another planet.

Reaching Out with Our Own Music

Beyond crewed flight, humankind has managed in its own way to musically reach beyond our solar system, with no greater ambassador to the depths of space than that of NASA's *Voyager* probes.

On 5 September 1977, a Titan IIIE/Centaur launch vehicle blasted off from Cape Canaveral in Florida. Onboard, *Voyager 1*, that was to make a

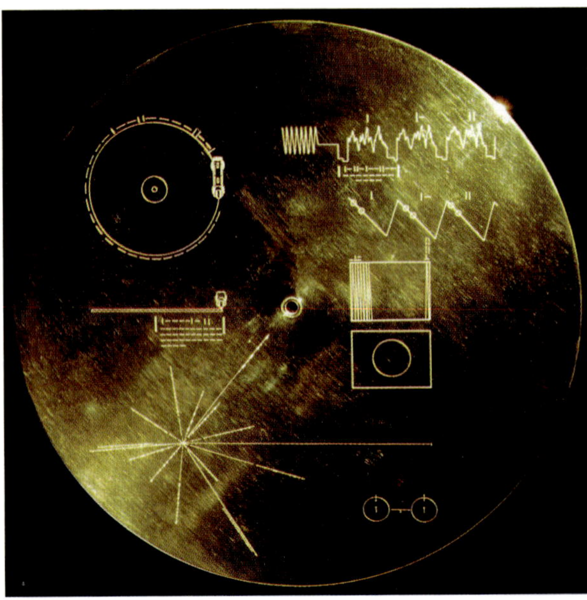

Described by cosmologist Carl Sagan as "… the launching of a bottle into the cosmic ocean …", a solitary golden record placed onboard NASA's *Voyager 1* and *Voyager 2* spacecraft, remain humankind's greatest emissaries to the stars. The discs that were sent forth in 1977 from Sagan's 'pale blue dot' contain 115 analogue-encoded photographs, greetings in 55 languages, and along with 12 minutes of 'Sounds of Earth', a further 90 minutes of music. (NASA/JPL)

journey that earned it the distinguished accolade of being the farthest travelled human-made object from Earth.

Voyager 1, and its sister craft *Voyager 2*, both carry Earth's very own 'message in a bottle', a Golden Record Disc attached to the outer body of the craft. The disc holds 115 images and Earth's very own entry into the Noisy Universe, a collection of natural sounds from the planet ranging from surf to that of wind and thunder. Other sounds include birds, whales, and a variety of wildlife to represent the diversity of life of on our planet. There is also a selection of music that lasts 90 minutes, Eastern and Western classics, and a variety of ethnic selections, and of course a greeting from humankind, purveyed in spoken word from Earth-people in 55 languages.

The greeting messages recorded on the disc start with *Akkadian*, the extinct language of the Akkad, written in cuneiform (wedge-shaped marks on clay tablets). It has two dialects, Assyrian and Babylonian, widely used from about 3500 BCE. It is the oldest Semitic language for which records exist. The greetings end with *Wu*, a modern Chinese dialect of which in 2007, there were 80 million native speakers.

The Golden Record Disc which is protected in an aluminium sleeve, complete with a cartridge and a needle, also has instructions in symbolic

The opening scene of Stanley Kubrick's science fiction masterpiece is enhanced by the dramatic sound of the timpani drums as they evoke a gloriously tantalizing array of imagery that spans the dawn of the prehistoric era through to humankind's conquest of space. The screenplay, written by Kubrick and Arthur C. Clarke, is powerfully brought to life by the use of Richard Strauss's stirring *Also Sprach Zarathustra*. (iStock/Furtseff)

language, explaining the origin of the spacecraft accompanied with information on how to play the record.

Along with Chuck Berry, the discs boast a vast musical spectrum, with Bach, Mozart, and even a blast on the bagpipes, plus a variety of different music to reflect the cultures from around the world.

Whilst most of the instruments onboard *Voyager 1* have been disabled, the craft entered interstellar space on 25 August 2012, becoming the first human-made object ever to do so. Along with its sister craft *Voyager 2*, whose route outward is different to *Voyager 1*, the pair conducted an epic 'Grand Tour' of the solar system's giant planets, flying by Jupiter, Saturn, Uranus and Neptune.

Gently and with an eerie silence, we can envisage *Voyager* gliding through space with a dark curtain backdrop scattered with countless stars, just as one would throw a handful of sand in the wind and watch the grains fall back to a deserted beach. Accompanying it in our minds, what better score than that of Stanley Kubrick's *2001: A Space Odyssey*, and Richard Strauss's *Also sprach Zarathustra*. Our very own spoken word and music mingles in the icy chill of space, probably never to be found, never to be heard, and never to be replied to.

We have taken and sent music into the universe, but long before humankind evolved on Earth, musical harmony and synchronicity, forged between the stars and other worlds, already existed and continues to do so.

Setting Sail on New Seas

The Cultural Significance of the Coming of the Space Age

Peter Rea

The Age of Sail to the Age of Spaceflight

He sat on the bank of the river. Somewhere upstream a tree had collapsed and was floating downstream toward the sea. He noticed a bird sitting on the log and a thought came into his head. After millions of years of evolution the enlarged human brain allowed him to think in abstract terms and be creative. This *Hominid*, later to be identified as *Homo Sapiens* (Wise Man) imagined himself sitting on the log and being taken down river. Later *Homo Sapiens*

The Outrigger Canoe was an early form of transport, which eventually led to sailing and steam power ships. (Wikimedia Commons/British Museum/Unknown Photographer)

would think about sitting inside a hollow log rather than sitting on it. Using primitive tools made of stone, flint, and deer antlers, the log could be hollowed out, making it easier to use. This enlarged brain, separating *Homo Sapiens* from other *Hominids* reasoned that a smaller log on the end of two branches attached to the main log would make it more stable after countless attempts to climb into the log resulted in falling into the sea or river. The outrigger canoe was invented. It is not possible to say who first imagined using a shaped piece of wood pulled through the water to propel the boat along. The next major leap of thought involved using an animal skin or woven reed mats stretched between poles to let the wind push the boat along. It was a major leap of thinking that would open the seas and oceans for explorers to look beyond the horizon in search of new lands. Humans really were setting sail on new seas. It would be the first steps down through thousands of years of time that lead to the Sea of Tranquillity and the first human footprints on the Moon.

The Beginning of the Space Age

On 5 March 1946 Winston Churchill gave a speech at Westminster College in the town of Fulton, Missouri in the USA. One famous line would go down in history; "From Stettin in the Baltic to Trieste in the Adriatic, an iron curtain has descended across the continent." The Cold War had begun. From this point on the American mass media and the public referred to the Russians as "Reds" or just "Communists." Whatever word they chose they were looked on as the enemy. In this background the USA were taken by surprise and shocked when, on Friday 4 October 1957, the Soviet Union became the first nation to launch an artificial satellite: Sputnik 1, a word meaning "Fellow Traveller."

A replica of Sputnik 1, the world's first artificial satellite. The replica is stored at National Air and Space Museum, Washington DC. (NSSDC/NASA)

Although only five at the time, I still have faint (and getting fainter) memories of this. It certainly got the attention of the world. The American press had a field day producing such sensationalist headlines as "Red made satellite flashes across US" and "Reds claim victory in satellite race." The American TV reported that Sputnik 1 was transmitting its characteristic bleep, bleep, bleep on 20 megacycles (now referred to as 20 megahertz). Radio hams were encouraged to listen out for it. TV news coverage at the time showed many people in the streets looking up at the heavens for signs of this new "star" crossing the United States. The TV news showed a map of the USA showing what states Sputnik 1 would fly over and when. A TV news crew went out onto the streets to take the public reaction. These can still be viewed on YouTube. When asked for her reaction one lady said, "It gets the American people alarmed that a foreign country, especially an enemy country, can do this." Another man said, "Definitely disturbing." The TV anchor said, "What secrets from the universe are the communists learning that we in the free world do not yet know." These sources capture the anxiety felt by the Americans in the wake of Sputnik 1. In contrast the English science fiction author Arthur C. Clarke would say of the event "The launch of Sputnik is one of the greatest scientific moments in world history." He saw no threat, only a new chapter in human history.

The USA would launch their first artificial satellite, Explorer 1 on 31 January 1958. An earlier attempt by Vanguard TV3 on 6 December 1957 ended in the rocket exploding on the launch pad causing one newspaper to call this mission "Kaputnik" another paper called it "Flopnik." The shock of Sputnik had not dampened the wit of the headline writers. I doubt whether any of those looking up at the night sky on the evening of 4 October 1957 would appreciate what a profound effect space technology would have on the world from this point forward. Those now living today use it daily, perhaps without knowing it. The Space Age had arrived.

How Spaceflight Affects All of Us: Communication

It became clear that a satellite could overfly all the countries of the world and both polar regions. Military minds saw potential. In the wake of WW2 and deteriorating relations with Russia, the USA was flying reconnaissance missions over Russia. In 1960 a U2 spy plane flying at an altitude of around 20 kilometres was shot down by a Soviet missile. Spying on the enemy was becoming a dangerous game. Satellites however were orbiting the Earth above 160 kilometres and way beyond the range of missiles at the time. It would not be long before both nations were putting high resolution cameras in their satellites and pointing them toward the ground. The reconnaissance satellite was born.

A slight detour is needed now before we move onto how satellites revolutionised worldwide communications for us all. I need the help of Johannes Kepler (1571–1630). His three laws of planetary motion explain how the planets move around the Sun. They also explain how satellites move around the Earth.

Law 1: There are no circular orbits: the orbit of a planet is an ellipse, with the Sun at one of the two foci. The effects of the Sun's gravity and other planets always make them an ellipse. In the case of satellites, it is the Earth at one foci.

Law 2: A line segment joining a planet with the Sun, sweeps out equal areas in equal time. Put simply, as a planet approaches the Sun it speeds up. When the planet is at its furthest from the Sun it is moving at its slowest.

Law 3: The square of a planet's orbital period is proportional to the cube of the length of the semi-major axis of its orbit.

This is the one that makes communication satellites a reality. As you move further from the Sun in the case of planets or Earth in the case of satellites, it takes longer to orbit the primary body. Now here is the clever bit. If a satellite is orbiting close to the Earth it will be moving at its fastest. The

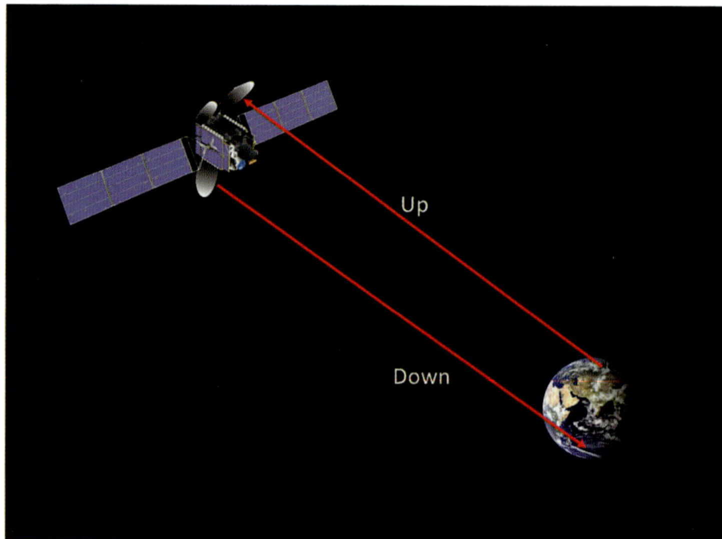

Travelling at an altitude of 35,786 kilometres above the equator, the satellite has an orbital velocity of 10,800 kilometres per hour, resulting in its orbital motion matching the Earth's rotation. From a given point on the Earth's surface the satellite would appear not to move and act as an orbital relay. (Earth and satellite images, NASA; diagram and captions, Peter Rea)

Space Shuttle orbiting around 200 kilometres high would have a velocity of 27,000 kilometres per hour, which is pretty fast. The Moon on the other hand at an average distance of 384,000 kilometres has an orbital velocity of 3683 kilometres per hour. There are of course intermediates of this. As you orbit closer and closer to the Earth, your orbital velocity will increase. If the satellite orbits at 35,786 kilometres, then thanks to Mr Kepler and orbital mechanics the satellite has an orbital velocity of 10,800 kilometres per hour and therefore its orbital motion matches the Earth's rotation. From a given point on the Earth's surface the satellite would appear not to move and act as an orbital relay for telephone, TV or Internet. Signals transmitted from a ground station cannot reach all parts of the Earth. From this vantage point, one tenth the way to the Moon, a communications satellite can view one whole hemisphere. Australia is on the other side of the world and the signal would be blocked by the bulk of the Earth, so it may need two satellites to make the connection. The signal journey would look like the letter M. That is up, down, up, down. My daughter lives in Australia and there is often a slight delay. I suspect but cannot verify that my voice follows that M pattern. Being able to communicate to most parts of the world has certainly revolutionised communications between countries and individuals. People can pick up their mobile (cellular) phone and send a text, send a picture or video. I "see" my daughter in Australia most weeks and keep up to date via a messenger service. All possible because of a network of satellites placed strategically above the equator at various longitudes. Something that was not possible before the space age.

How Spaceflight Affects All of Us: Navigation

I like to think I have a reasonable sense of direction. Being born in the northern hemisphere I am acutely aware that the Sun rises in the east, culminates (reaches its highest point in the sky) in the south and sets in the west. To be pedantic it only rises due east and sets due west at the equinoxes. At other times it will rise north of east in summer and south of east in winter. The same applies to setting. If I am in an unfamiliar place, I always know where south is by noting the position of the Sun and knowing the time of day. On a cloudy day I only need to look at satellite dishes on people's houses. They always point toward the south in the northern hemisphere. This worked fine until I visited my daughter in Australia. I became disorientated because whilst the Sun rises in the east, it culminates in the north and satellite dishes are also pointing north. My built-in homing pigeon ability was now out by 180 degrees! Satellite navigation, a relatively recent use of satellite technology has great cultural significance, especially for those folks lacking any sense of direction. The USA was the first to have an operational satellite navigation system, the Global Positioning System, or GPS. Other countries followed later. The GPS system uses a "constellation" of satellites in different orbital planes or angle to the equator at an altitude of 20,350 kilometres. This ensures that at least four satellites are above the horizon from any given location. GPS satellites broadcast radio signals providing their locations, and precise time from on-board atomic clocks. The GPS radio signals travel down to us at the speed of light 299,792 kilometres per second. A GPS device receives the radio signals, noting their exact time of arrival and uses these to calculate its distance from each satellite in view. Once a GPS device knows its distance from at least four satellites, it can use geometry to determine its location on Earth in three dimensions. These receivers used to be much larger than the ones we use now. My car has one built in and so does my iPhone. No excuse for getting lost if my sense of direction fails me. There are other uses for satellite navigation other

The GPS navigation system uses a "constellation" of satellites in different orbital planes at an altitude of 20,350 kilometres. This ensures that at least four satellites are above the horizon from any given location at any one time, allowing you to precisely fix your position. (USAF)

than the obvious one of taken us from A to B. In an emergency they can give your precise position and height above sea level for rescue services. The system is used to improve maps and by setting up stations along geological faults, crustal movement can be determined, which improves earthquake predictions. GPS has changed our lives for the better I believe. It has changed our modern culture to rely on technology. Yet despite this I still carry an up to date road atlas in the car.

How Spaceflight Affects All of Us: Looking After Planet Earth

From that special vantage point 35,786 kilometres above the equator satellites fitted with cameras and other instruments can view the whole of one hemisphere. This enables the satellite to take pictures of the Earth, at the same location, every thirty minutes. Computer processing of this data creates "movie loops" which shows weather systems developing and their movements can be tracked to give more accurate forecasts. In periods of extreme weather the number of images taken can go from one every thirty minutes to one every five minutes or less. The television news and weather programmes make their own "movies" for us to see every day. Will I need an umbrella today? Watch the weather forecast or app on your smart phone and you can get accurate (mostly) forecasts of when those rain bearing clouds will arrive. Weather satellites and other Earth monitoring satellites can also be placed into a near-polar orbit around the Earth. These near-polar orbiting satellites commonly choose a Sun-synchronous orbit, where each successive orbital pass occurs at the same local time of day. The altitude of these satellites is usually between 700 and 800 kilometres, giving much higher resolution than geostationary satellites. From this vantage point they can keep an eye on the health of our home planet.

Another slight detour is needed at this stage to make a point about a change in culture relating to the Earth itself. With our new capabilities to explore space and the planets in particular, the early Mariner missions from the USA showed that two planets, Venus and Mars, had undergone significant changes. Venus is covered in a dense carbon dioxide (CO_2) atmosphere caused by what scientists called a runaway greenhouse effect. The Sun's energy passed through the atmosphere but on hitting the ground was radiated back up through the atmosphere at infra-red wavelengths. The carbon dioxide trapped this heat causing the planet to warm up. It is hot enough at the surface of Venus to melt lead. The atmosphere of Venus is also very acidic. Not a nice place to be and hostile to spacecraft wanting to descend through the atmosphere to the surface. Mars on the other only has a very thin atmosphere. There is ample evidence that in the past the atmosphere was dense enough for free-running

Venus is covered in a dense atmosphere, while most of the Martian atmosphere has gone. Both scenarios are due to climate change taking place over billions of years. (Venus image NASA/JPL-Caltech; Mars image NASA/Lošmi; captions Peter Rea)

water to cover the surface of Mars. Over time the atmosphere eroded away, and the water disappeared. Mars is no longer a warm wet planet but a cold dry one. Both Venus and Mars suffered significant climate change. Earth does too. The last Ice Age came to an end about 10,000 years ago. Water levels have risen as the polar ice melted. Not far from where I live there is a U-shaped valley, and outside a local library is a large "erratic" boulder made of rock that is not native to this region. Both are evidence of glaciation in the recent past, on the geological time scale. The U-shaped valley is carved out by a glacier and the rock was transported by one. In just my lifetime we have become aware that human actions like pumping smoke and carbon dioxide into the air from our factories and coal-fired power stations are also contributing to a changing climate. Within my memory I recall the London smogs that made it difficult to breathe if you had respiratory issues. Satellite monitoring of our planet has done much to make us all aware of this and other climate issues like deforestation and the build-up of CO_2 and methane in the atmosphere, both greenhouse gases. Climate change is a hot topic these days and rightly so. Earth monitoring satellites are giving us the data and evidence to make informed decisions about the future. Let us use that information wisely.

Taken from the International Space Station, this view of the Earth below shows the thin line of the Earth's atmosphere. (NASA)

Culturally the space age has shown all of us that the Earth is a fragile planet. Viewed from afar it looks like a pale blue dot. Closer in, we can see that the Earth's atmosphere looks very thin and cannot keep soaking up our pollution. Something must change. I believe it is, albeit slowly. Space age technology helps us monitor and take the pulse of our planet, and in so doing changes our culture from one of abusing our planet to one of caring for it.

Will our Culture Survive?
Throughout geological time there have been many catastrophes that have led to mass extinctions. The one occurring around 66 million years ago killed off 75% of all species including the non-avian dinosaurs. More than 99 percent of all organisms that have ever lived on Earth are extinct. We may be in a mass extinction now. It has a name, the Holocene extinction, with flora and fauna becoming extinct at hundreds of times the background rate over the last century. Are humans at risk? All species are. If humans became extinct during the current century, how will we be remembered? The buildings and structures of our genius will collapse into dust in a few thousand years and in time there will be no evidence that *Homo Sapiens* ever existed, at least on Earth. Future explorers from other planets within the galaxy could find evidence of our existence. During the first six decades of Earth's space age, human artefacts have been left on other planets within our solar system and

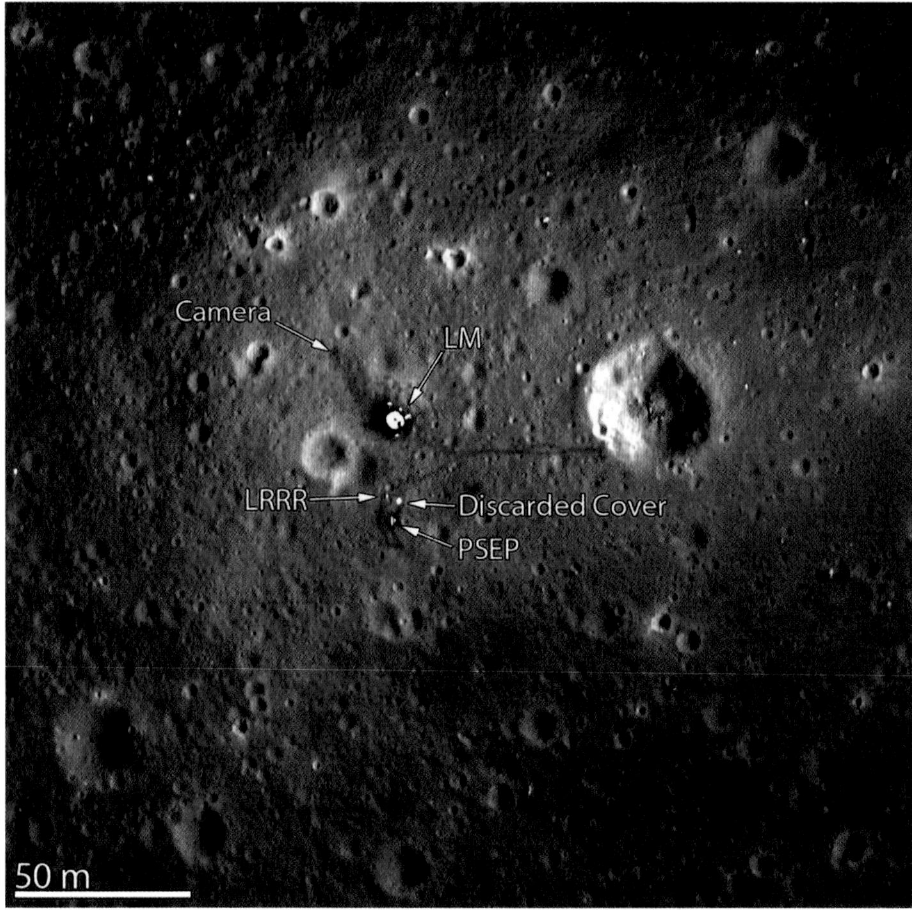

This image of the Apollo 11 landing site and descent stage captured from just 24 kilometres (15 miles) above the surface provides Lunar Reconnaissance Orbiters' best look yet at humanity's first venture to another world. Future visitors from Earth may well look in awe at what humans achieved in the 1960s. Future visitors from another planet may well assume that inhabitants of the nearby planet above the horizon once came this way. (NASA Goddard/ Arizona State University)

five spacecraft are heading out into the galaxy. Other human artefacts could be found orbiting the Earth. Low earth satellites will have decayed and burned up in the atmosphere. Communications satellites in geostationary orbit one tenth of the way to the Moon will stay in orbit, silent and dead, indefinitely.

The Moon has spacecraft from the Soviets, the Americans, and the Chinese, including some wheeled rovers. The descent stages from six Apollo landings still sit on the Moon's surface along with three Lunar Roving Vehicles that carried six humans across the lunar surface. Also left on the Moon are six science stations placed there by the Apollo astronauts who also left personal items of cultural significance. Apollo 14 astronaut Alan Shepard being an avid golfer took a golf ball to the Moon and using a special attachment shaped like the head of a golf club fitted to a tool that had been used to pick up lunar rocks he hit the golf ball for "miles and miles". What any future explorers from another planet would make of this golf ball is anyone's guess. They might be surprised to find out that a long time ago a sentient species capable of spaceflight found great pleasure in trying to hit this ball into a hole using a metal rod with a handle at one end and a flattened piece of metal at the other end. They would not appreciate that the only culturally significant ball game is, of course, cricket! The Apollo 15 crew left behind a small model of an astronaut in a space suit representing a "fallen astronaut" and a plaque giving

On their final moonwalk, the Apollo 15 astronauts David Scott and Jim Irwin created a small memorial to those astronauts and cosmonauts known to have lost their lives whilst on active duty. This very act shows that human culture has evolved compassion and feelings for the dead, something that is not exclusively human as other animals have shown similar traits. (NASA)

Charlie Duke's family portrait left on the surface of the moon. Cat Crater at Station 14 was named for sons Charles And Tom. Dot Crater at Station 16 was named for his wife. (NASA)

the names of the 14 astronauts and cosmonauts who had died whilst serving. Apollo 16 astronaut Charlie Duke left behind a family portrait.

There are many other human artefacts left on some of the planets. Venus has numerous soviet spacecraft on its surface whilst Mars has a retinue of stationary landers and rovers. Even the largest moon of Saturn – Titan – has an artefact from Earth in the form of the European Space Agency's Huygens lander. Finding these artefacts relies on other sentient and spacefaring species visiting our solar system and locating that golf ball. In an initiative-taking move NASA has sent five spacecraft on solar system escape trajectories, meaning they have sufficient velocity (energy) that the Sun's gravity will not pull them back and they will leave the solar system to wander the galaxy. In order of launch they are *Pioneer 10*, *Pioneer 11*, *Voyager 2*, *Voyager 1*, and *New Horizons*. (Yes, Voyager 2 was launched before Voyager 1.) On the remote chance that these spacecraft will be found, all of them were fitted with various means to tell any inhabitants of the galaxy something about us. New Horizons was fitted with a compact disc (CD) containing 434,738 names of some inhabitants of Earth. Mine is one of them and it is well into the Kuiper Belt by now. The Voyagers had a gold record fitted with many examples of our culture: music, pictures, sounds and various greetings. This is discussed in the chapter 'Music and the Cosmos' elsewhere in this book.

The plaques on the Pioneer are worthy of some discussion. Their message is not so much on culture, as with the Voyager gold record, but a statement of where we are and that a sentient, spacefaring species sent it out there. The plaques are made of aluminium which has been gold anodised. They measure 229mm × 152mm and have the messages engraved onto the surface. On the right hand side can be seen the outline of the Pioneer spacecraft with a naked man and woman shown in front of it for scale. In a slightly controversial way in keeping with the period the man's genitalia are shown but not the woman's. It was an innocent way of depicting that there are two sexes required to continue our species as is most, but not all, life on Earth.

At the top left of the plaque is a diagram showing the hydrogen atom, which is the simplest and most common element pervading the cosmos. Our astronomers using optical, or radio telescopes use a region of the spectrum

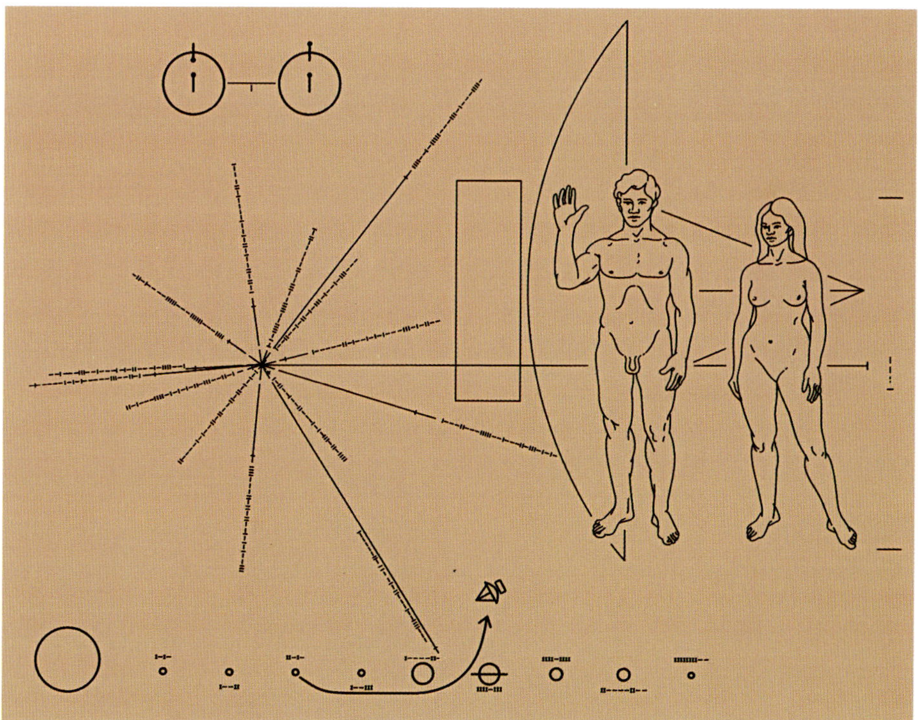

Like a message in a bottle the Pioneer plaque – and the gold record on both Voyager spacecraft – sends a greeting to any advanced civilisation lucky enough to find them that we exist, this is where we live, and here is our culture. It is a way of preserving our species, our culture and our memories of *Homo Sapiens*, the first spacefaring species of planet Earth. (NASA/Oona Räisänen (Mysid)/Carl Sagan/Frank Drake/Linda Salzman Sagan)

called the 21 centimetre line. This comes from a property of the sole electron of the hydrogen atom called "spin" which can flip from two states of "spin" in a very short time, so radiating energy at this 21 centimetre wavelength. As the hydrogen atom would be known to other sentient species it is used on the plaque as a standard unit of measurement and time.

Just below the atom diagram is the binary unit 1 which in this case equates to the 21 centimetre wavelength. To the right of the woman is the binary unit 1-0-0-0 or 8 in base 10. Multiplying 8×21 equates to 168 (centimetres), this being a typical height for a human female. I had to have this explained to me, so I hope another civilisation will be able to decipher it as intended by the plaque designers.

Below the hydrogen atom diagram is another diagram that shows the location of 14 pulsars with respect to our Sun. The binary digits on each line indicate the period of the pulsar. The unit of time is taken from the time taken for the hydrogen electron to flip its state from up to down or vice versa. Deciphering this will indicate our location in the galaxy.

The final diagram is of our solar system which shows Pioneer being sent from the third planet and passing the fifth planet (Jupiter) before leaving the solar system. After Pioneer 11 had passed Jupiter, it was redirected for an encounter with Saturn. As both plaques are identical and fitted before launch the Saturn encounter is not shown on the Pioneer 11 plaque.

Humans from Earth have therefore left artefacts on the Moon as well as on other planets, and sent (so far) five products of our genius out into the galaxy in the hope that some of our culture will survive, even if we don't, and perhaps for more advanced spacefaring cultures to consider us worthy of contact and mutual interaction. What a legacy for our own infant spacefaring species that would be.

We Set Sail on This New Sea
The title to this chapter comes from a speech given by the late US President John F. Kennedy at Rice University in Texas on 12 September 1962. He used this opportunity to seek support for his goal of putting a man on the Moon before the end of the decade. "We set sail on this new sea because there is new knowledge to be gained, and new rights to be won, and they must be won and used for the progress of all people."

The space age has changed our perception of space travel from one of science fiction to one of science fact. Will it push us to explore further than before? Humans have always been explorers and innovators, first setting out in canoes or rafts, and later using the power of wind with sailing ships. They harnessed the power of steam to propel their ships, and subsequently used the power of

President John F. Kennedy's address at Rice University, Houston, Texas, concerning the nation's efforts in space exploration. In his speech the President discusses the necessity for the United States to become an international leader in space exploration and famously states, "We set sail on this new sea because there is new knowledge to be gained, and new rights to be won, and they must be won and used for the progress of all people." (Wikimedia Commons/Robert Knudsen)

the atom to do the same. Having explored this planet, where do we go next to sate our desire to explore? The Moon? Mars? Perhaps the moons of Jupiter? Can we ever leave our solar system and set out for the stars? If so, would we be the first to do this, or is the galaxy teeming with space-faring species?

I do not know the answers to these questions. What I do know is that the space age changed our culture and the way we think about our planet ... and it all began on 4 October 1957.

Full Moon Names

A Full Moon by Any Other Name Would Shine as Bright

Lynne Marie Stockman

Every month, media outlets around the world announce the arrival of the upcoming Full Moon, with details of how, where and when to see it. And it is inevitably bestowed with a name, suitably translated into the local language, which usually is drawn from the list below.

Month	Full Moon Name	Meaning	Origin
January	Wolf	Hungry wolves are howling	Celtic
February	Snow	Heavy snow is falling	Celtic
March	Worm	Worms are emerging from the thawing ground	Native American
April	Pink	Pink phlox are blooming	Native American
May	Flower	Plants are flowering	Many cultures
June	Strawberry	Strawberries are ready for harvesting	Native American
July	Buck	Deer are growing antlers	Native American
August	Sturgeon	Sturgeon are appearing in great numbers	Native American
September	Harvest	Crops are ready for harvesting	Many cultures
October	Hunter's	Preparing for winter by hunting for food	Native American
November	Beaver	Beavers are preparing for winter	Native American
December	Cold	Winter is here	Celtic

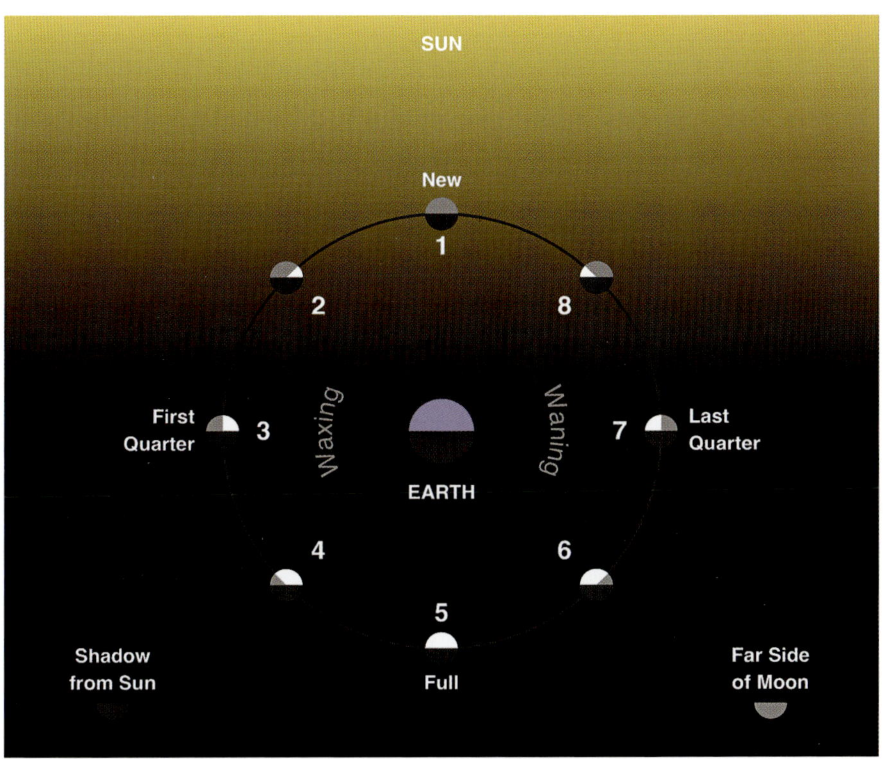

Northern Hemisphere

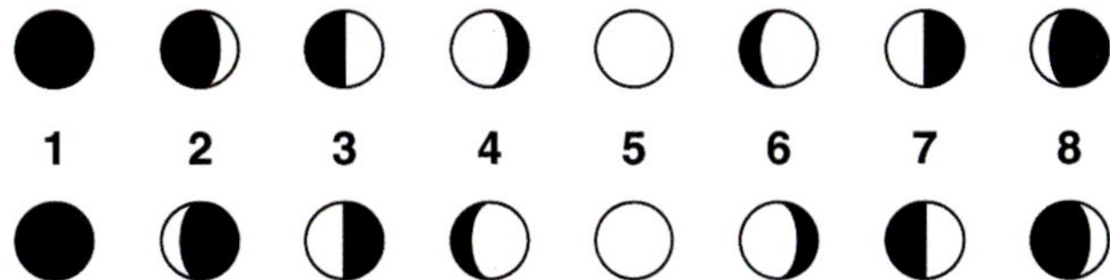

Southern Hemisphere

A Moon phase is simply the shape of the lit portion of the Moon. A lunation typically begins with New Moon (which is not visible except during a solar eclipse), followed by the waxing crescent Moon (less than half illuminated), First Quarter Moon (half illuminated), and waxing gibbous Moon (more than half but not fully illuminated). Full Moon occurs at the instant when the ecliptic longitudes of the Moon and Sun differ by exactly 180°. The waning phases follow: gibbous, Last Quarter, crescent. The observer's location on Earth is also significant as to which part of the Moon appears lit. (Lynne Marie Stockman)

What's in a Name?

Волчья Луна: каким было первое полнолуние 2022 года

(Дзеркало тижня, 18 January 2022)[1]

---

1. Ukranian: Wolf Moon: what was the first full moon of 2022 like?

Was steckt hinter dem „Schneemond"-Spektakel?

(Die Welt, 9 February 2017)[2]

Oblohu rozzářil poslední letošní superúplněk, „červí Měsíc" ukončil zimu

(Mladá fronta DNES, 21 March 2019)[3]

Esta noite tem uma boa razão para ir à janela e olhar para o céu. Vem aí uma lua 'cor-de-rosa'

(Correio da Manhã, 7 April 2020)[4]

Dünya merakla bekliyor… "Süper Çiçek Kanlı Ay" tutulması yarın gece gerçekleşecek!

(Hürriyet, 25 May 2021)[5]

Truskawkowy księżyc w pełni już wkrótce na niebie.

(Gazeta Wyborcza, 13 June 2022)[6]

«Φεγγάρι του Ελαφιού»: Εντυπωσιακές εικόνες από την υπερπανσέληνο σε όλο τον κόσμο

(Η Καθημερινή, 14 July 2022)[7]

Super Lune de l'Esturgeon: Toutes les informations à savoir pour profiter pleinement de la vue ce soir.

(Journal L'union, 11 August 2022)[8]

Luna de cosecha: por qué se llama así, cuándo aparecerá y cómo verla

(La Nación, 22 September 2022)[9]

Starwatch: moonlight bright enough to hunt by.

(The Guardian, 21 October 2018)

Luna plină din data de 19 noiembrie mai este cunoscută și ca Luna Castor, fiind denumită așa în Statele Unite.

(Evenimentul Zilei, 13 November 2021)[10]

---

2. German: What is behind the "Snow Moon" Spectacle?
3. Czech: The last super full moon of the year lit up the sky, the "worm moon" ended winter.
4. Portuguese: Tonight you have a good reason to go to the window and look at the sky. A 'pink' moon is coming.
5. Turkish: The world is waiting in anticipation… The 'Super Flower Blood Moon' eclipse will take place tomorrow night!
6. Polish: The strawberry full moon will soon be in the sky.
7. Greek: "Buck Moon": Impressive images of the super full moon around the world.
8. French: Sturgeon Super Moon: All the information you need to know to fully enjoy the view tonight.
9. Spanish: Harvest Moon: why it is called that, when it will appear and how to see it.
10. Romanian: The full moon on 19 November is also known as the Beaver Moon, as it is called in the United States.

Ecco le immagini scattate nel mondo di quella che prende il nome di "Luna fredda"…

(la Republica, 14 December 2015)[11]

However, there is a rich and varied set of traditions amongst the indigenous peoples of North America, not to mention names imported by European colonists to the New World, which provide many interesting alternatives to the names given previously.

Winter (December–February)
The names of the winter Full Moons inevitably invoke sensations of cold and visions of darkness. Freezing conditions can lead to tree sap expanding and eventually exploding outwards through the bark. The Cree, Northern Arapaho, Oglala and Western Abenaki tribes are but a few that name one of the winter Full Moons after this phenomenon. For the Lakota in the upper Mississippi River and Great Plains regions, the January/February Full Moon is *Cannápopa Wi*, the *Moon When Trees Crack From the Cold*. Other 'cold' names include the *Freeze Up Moon* (Algonquin), the *Frost Sparkling in the Sun Moon* (Northern Arapaho), and the *Someone's Ears are Freezing Moon* (Oneida).

The winter solstice of the Northern Hemisphere occurs in mid-December; this leads to short days and long nights. The Inupiat living in Alaska above the Arctic Circle name the December Full Moon *Sikiñgilak*, the *No Sunshine Moon*. For the Mohican people, the Full Moon nearest to the solstice is the *Long Night Moon* but for the Zuni of New Mexico, who recognise that the December solstice Sun has reached its southernmost declination, this Full Moon is *Ik'ohbu Yachunne*, the *Sun Has Travelled Home to Rest Moon*. The Germanic festival of Yule, celebrated around the time of the December solstice, features in Anglo-Saxon Moon names: the Full Moon before solstice is the *Moon Before Yule* and the Full Moon afterwards is the *Moon After Yule*. *Wicogandu*, the *Centre Moon* of the Assiniboine, takes place in January, midway through the winter season on the northern Great Plains. For Colonial Americans, the Full Moon nearest to Christmas was the *Christmas Moon* and the following one was the *Winter Moon*.

During the darker months of the year, the spirit world may feel particularly close. The Ojibwe of Southern Canada designate the January Full Moon *Gichimanidoo-giizis* or *Great Spirit Moon*. For the Northern Arapaho, this same moon is the evocatively named *When the Snow Blows Like Spirits in the Wind Moon*. In the south eastern woodlands of the United States during February, *Kagali* is the *Bony Moon* of the Cherokee, and marks the time when a family meal is prepared, complete with places set for departed family members.

---

11. Italian: Here are the images taken in the world of what is called the "cold moon" …

The Full Moon of 21 March 2019 appears behind the Wenatchee Mountains of north central Washington in the United States. This particular Full Moon was the third 'largest' (biggest apparent diameter) of the year and considered by some to be a 'Supermoon'. (Frank Cone/Pexels)

Spring (March–May)

The snow pack is at its greatest extent in late winter and early spring, and the Sun's bright rays rebounding off of the snow can lead to snow blindness. The *Moon of Sore Eyes* appears in the traditions of a number of North American tribes, including the Assiniboine who dub it *Wicinstayazan*. Full Moons named after hard-crusted snow, damaged snowshoes, and rivers of broken ice appear in March as winter gives way to spring.

Spring means rebirth, so most of the names of the Full Moons during this season reflect the return of light and warmth and life to a frozen land. Many tribes celebrate the return of various bird species – crow, duck, eagle, goose, loon – by naming Full Moons after them. In the north eastern woodlands of Canada and the United States, the Western Abenaki celebrate them all with *Sigwankas*, the *Birds Returning Moon*. The budding and blooming of plants is also noted. For the Tlingit clans along the Pacific coast of British Columbia and Alaska, the March Full Moon is *Héentáanáx̱ Kayaan'i Dís*, the *Underwater Plants Sprout Moon*, and for the Ojibwe, this same Full Moon is named *Ziinsibaakwadooke-giizis* or the *Sugar Maker Moon* after the freely-running sap in the trees. Many of the Native American Full Moon names revolve around plants and berries and springtime activities: *Blossom Moon, Budding Moon, Earth Moon, Egg Laying Moon, Flower Moon, Frog Moon, Moon of Shedding Ponies, Moon of the Red Grass Appearing, Moon When the Leaves are Green, Moose Hunter Moon, Planting Moon*, and so on. Blue camas (Camassia quamash) is a vital food source of the Kalapuya people situated in Oregon; four of their spring and summer Full Moons – *Women Dig Camas Moon, Time for Pounding Camas Moon, Camas Blooming Time Moon*, and *Camas Ripe Moon* – are named for this all-important plant.

Anglo-Saxon and Celtic traditions for spring Full Moons include the *Plough Moon* (March), *Egg Moon, New Shoots Moon* and *Seed Moon* (April), and *Milk Moon* (May). The unpredictable nature of the weather is noted in the naming of the *Wind Moon, Strong Wind Moon*, and *Windy Moon* by cultures on both sides of the Atlantic.

Summer (June–August)

The Full Moon names of summer celebrate the ripening of crops and the arrival of hot weather. A number of Full Moons across North America bear the names of black cherries, blackberries, blueberries, chokeberries, plums, squash, and strawberries. The Assiniboine economically designate one of their summer Full Moons as *Wasasa*, the *Red Berry Moon*, and in late summer, the Tlingit commemorate *Sha-ha-yi*, the *Berries Ripe on the Mountain Moon*. For the Zuni, heavily-laden fruit trees are remembered with *Dayamcho Yachunne* or the *Moon When Limbs of Trees are Broken by Fruit*.

The cultivation of crops is a major summertime activity and Full Moon names reflect this. Amongst the Western Abenaki, *Nokahigas* or the *Hoer Moon* in early summer lights the evening when hoes are utilised to break up the soil and remove weeds. Maize is an important crop: the Cherokee have the *Green Corn Moon* and *Ripe Corn Moon* in June and July whilst the Algonquin name no fewer than five of their Full Moons after various aspects of corn cultivation. The cutting of wild rice, grass, or hay is an important endeavour in

July, with the Anglo-Saxon *Hay Moon* occurring in the middle of season. The English colonists of North America, however, would take time to smell the roses under the June *Rose Moon*.

Salmon is an important resource along the northern Pacific Coast and both the Tlingit and their neighbours, the Haida, name summer Full Moons *X̱aat Dísi* and *Chíin Kungáay*, respectively, after this fish. Other Full Moons of summer mark the moulting of bird feathers and the loss of velvet on a caribou's antlers. The arrival of newborn animals is also represented, with the Celtic *Horse Moon*, Cree *Hatching Moon*, Inupiat *Birthing Moon*, and Tlingit *Birth Moon* taking place in June.

The warm and often stormy weather of summer gives rise to the *Hot Moon* and *Summer Moon* of various cultures as well as *Hiyeswa Tiriri Nuti* or *Thunderstorm Moon* of the Catawba people in the Carolinas.

Autumn (September–November)

Autumn is a period of anticipation and preparation. Crops are harvested, leaves fall, and the weather turns cold. The September/October Full Moon is associated with the harvest across North America and Northern Europe: *Barley Moon, Corn Moon, Gourd Moon, Mulberry Moon, Nut Moon, Rice Moon,* and so on. Food stuffs are stored away and the hunt begins in earnest to provide ample protein over the cold, dark months of winter. In the middle Mississippi River region, the Tunica people call their October Full Moon *Tahch'awera*, the *Hunting Moon*.

The turning colour and falling of the leaves of deciduous plants takes place during this time. For the Lakota, *Canwapekasna Wi* is the *Moon When the Wind Shakes Off Leaves*, but many other tribes also name their September/October Full Moons after leaf-fall.

Autumn is the breeding season for many animals and marks the return of shaggy coats to ward off winter cold. For the Inupiat, the October Full Moon is *Nuliaġvik*, the *Mating Caribou Moon*, and the November Full Moon is *Waníyetu Wi*, the *Rutting Deer Moon*, of the Lakota. The Haida mark the hibernation of black bears with their Full Moon of October/November and the Cree watch the birds fly south.

The late autumn Full Moon names presage the approaching winter season. Freezing weather and early snow appears in many Full Moon names: *Freezing Moon, Freezing River Moon, Frost Moon, Snow Moon*. The Kalapuya sensibly name their November Full Moon *Alangitapi*, the *Moving Inside for Winter Moon*, whilst the Inupiat note the lengthening nights with *Nippivik*, the *Sunset Time Moon*, of late November. For those hardy souls living north of the Arctic Circle, the Sun will soon disappear below the horizon, not to rise again until well after the winter solstice.

The Full Moon of February 2021 rises above the Earth as seen from the International Space Station. (NASA/Mark Garcia)

Full Moon Traditions of China

The traditional Chinese calendar is lunisolar. Each month is either 29 or 30 days long and a year normally contains 12 such months. However, a 30-day intercalary or 'leap' month is inserted every 32 or 33 months to keep the lunar months in sync with the seasons. The Chinese New Year begins at the second or third New Moon following the winter solstice in December; this places it sometime between mid-January and mid-February. Since a month begins at New Moon, the Full Moon always occurs mid-month. The *Holiday Moon* is the first Full Moon of this calendar and is celebrated with the Lantern Festival (yuán xiāo jié or 元宵节) which marks the end of New Year festivities.

The *Budding Moon* is the second Full Moon of the traditional calendar with the Flower Festival (huā zhāo jié or 花朝节) taking place at this time. This celebrates the birthday of the flower goddess who, according to ancient belief, is thought to control human reproduction. Keeping her happy ensures future prosperity.

The following Full Moons are called the *Sleepy Moon*, *Peony Moon*, *Dragon Moon*, and *Lotus Moon*.

The seventh Full Moon in the lunisolar calendar is the *Hungry Ghost Moon*. On this day, the gates of the underworld are opened and ghosts and spirits return to roam the earth, seeking entertainment and nourishment. During the Ghost Festival (zhōng yuán jié or 中元节), family members offer prayers for their deceased relatives, burn special paper objects representing money and everyday items, and leave out food and drink.

The next Full Moon is the *Harvest Moon* which is the focal point of the Mid-Autumn Festival (zhōng qiū jié or 中秋节), a holiday second in importance only to the New Year. Lanterns, symbolic beacons lighting people's way to prosperity and good luck, are carried, and traditional treats like mooncakes are consumed. This is also a common time to celebrate courtship and marriage.

The Full Moon following this is the *Chrysanthemum Moon*. The last major Full Moon festival (xià yuán jié or 下元节), celebrating the birth of the Water Official in Taoist belief, takes place in the tenth month during the *Kindly Moon*. The final two Full Moons reflect the onset of winter, the *White Moon* and the *Bitter Moon*.

Similar festivals are celebrated on Full Moons throughout East Asia, including in Japan, Korea, Tibet, and Vietnam.

Buddhism and the Full Moon

Like the Chinese calendar, the Buddhist calendar is lunisolar, comprised of 29- and 30-day lunar months beginning with the New Moon and with the Full Moon falling on the fifteenth day of the month. Full Moons have particular religious significance to Buddhists because tradition holds that many of the important events connected with the Buddha – his birth, his enlightenment, his first sermon, his attainment of Nirvana, etc. – occurred at Full Moons.

In Sri Lanka, for instance, each Full Moon has a name which memorialises some significant aspect of the Buddha's life and teachings, and each Full Moon day (*poya*) is a public holiday in the country. *Duruthu Poya* is the first Full Moon in January and marks the Buddha's first visit to Sri Lanka. *Navam Poya*, typically in February but sometimes in late January, commemorates two important events, the appointment of the first two chief disciples of the Buddha and the meeting of the first Buddhist Council after the death of the Buddha. *Medin Poya* remembers when the newly-enlightened Buddha first visited his father. *Bak Poya*, in late March or April, marks the Buddha's second visit to Sri Lanka. *Vesak Poya*, one of the most important Buddhist festivals, is the birthday of Siddhartha Gautama (birth name of the Buddha) and is celebrated in May. *Poson Poya* follows in June and pays tribute to the conversion of Devanampiya Tissa, an early king of Sri Lanka, to Buddhism.

The Full Moon of July is *Esala Poya*. On this day the Buddha delivered his first sermon. *Nikini Poya* is next, marking the first convocation of Buddhist monks. This is followed in the next lunar cycle by *Binara Poya* which usually falls in September and commemorates the founding of the Order of the Female Buddhist Monastic. The next Full Moon day, *Vap Poya*, mostly occurs in October and salutes several important events in the life and works of the Buddha. *Il Poya* in November remembers the commissioning of 60 disciples by the Buddha himself and *Unduvap Poya* celebrates the arrival of a sapling taken from the tree under which the Buddha attained enlightenment. This tree remains one of the most sacred relics of Buddhism in Sri Lanka. If the year contains 13 rather than 12 Full Moons, then an additional *poya*, called an *adhi poya*, is inserted somewhere in the calendar to keep it synchronised with the Sun.

Modern Full Moon Traditions of South Africa

All of the seasonal names that we have seen thus far originate in the northern hemisphere. Because the seasons of the southern hemisphere are half a year out of step with the north, these Full Moon names do not make much sense south of the equator. Equally, plants and animals that live in North American, Europe and Asia may not exist in South America, Africa and Australia. The Centre for Astronomical Heritage in Cape Town, South Africa, is working to establish Full Moon names that are meaningful to the local population. Since it is possible to have two Full Moons in any month except February, 23 new names have been invented to cover every eventuality.

A few of the plants and animals inspiring names of Full Moons around the world. Photo credits (Wikipedia): Camas (William & Wilma Follette, United States Department of Agriculture), Chrysanthemum (Ronincmc, CC BY-SA 4.0), Corn (Sam Fentress, CC BY-SA 2.0), Salmon (United States Environmental Protection Agency), Meerkat (Jan Mehlich, CC BY-SA 2.0)

The *Mantis Moon* is the first Full Moon of January and honours the praying mantis, a central figure in the religion of the San people of South Africa. Should January have a second Full Moon, it is called the *Leopard Moon*. The February *Dassie Moon* is named for dassie or rock hyrax. The southern hemisphere autumnal equinox occurs in March, bringing the *Harvest Moon*, but if March has a second Full Moon, it is called the *Ochre Moon*. Ochre has been used for millennia by the indigenous peoples for personal adornment. South Africa is rich in mineral wealth, and the Full Moon(s) of April, the *Diamond Moon* (first) and the *Gold Moon* (second), recognise the importance of the mining industry. The *Frost Moon* of May announces the coming of winter with a possible second Full Moon called the *Fire Moon* as fires must be lit to keep warm at this time of year. The June *Sisters Moon* recognises the increasing visibility of the Pleiades (Seven Sisters) star cluster. If June has two Full Moons, the second is the *Honey Moon* since honey is an important nutritional supplement during the winter.

The humble meerkat is commemorated with the first Full Moon of July with a possible second instance called the *Protea Moon* after the national flower of South Africa. This is followed by the *Peace Moon* in August. Since wind storms are common at this time of year, any second August Full Moon is named the *Dusty Moon*. Spring returns to the southern hemisphere in September, hence the *Spring Moon* of that month. A second September Full Moon is named after the national bird of South Africa, the blue crane. The October

Full Moon(s) are named after whales and elephants, and November brings the *Milk Moon* and possibly the *Wool Moon*. South Africa's national animal, the springbok, adorns the name of the first December Full Moon, whilst the eland, the largest antelope in Africa, lends its name to any second Full Moon of the month.

Other Full Moon Names

Overlapping the spring Full Moon names is the *Paschal Moon*. This is the first Full Moon occurring after the vernal equinox and may take place in either March or April. The timing of this Full Moon determines the date of Easter, the most important day in the Christian religious calendar. The actual date of Easter relies on something called the Ecclesiastical Full Moon, a fictitious computed entity which does not correspond exactly to the actual astronomical Full Moon, but that's another story!

The modern *Harvest Moon* is the Full Moon nearest to the September equinox. This typically occurs in September, but can appear in early October. If either September or October has two Full Moons, then the three Full Moons covering these two months are the *Corn Moon*, the *Harvest Moon*, and the *Hunter's Moon*.

> Bijzondere blauwe supermaan in ons land goed te zien
>
> (*De Telegraaf*, 30 January 2018)[12]

A synodic month (the time it takes for the Moon to complete a cycle of phases) is about 29.5 days long. Because 12 synodic months are around 11 days shorter than a solar year, there is an extra Full Moon approximately every 2.5 years. As originally used in nineteenth-century farmers' almanacs, a *Blue Moon* is the third Full Moon in a season containing four. Inserting this extra Full Moon name allows the almanac makers to maintain the traditional Full Moon names in their proper months when a year has 13 rather than 12 Full Moons. However, a mistake in an article about Blue Moons in the astronomy magazine *Sky & Telescope* that was subsequently popularised by the board game *Trivial Pursuit* led to the alternative and now more common definition of a *Blue Moon* being the second Full Moon in a calendar month.

> 'Blood Moon' total lunar eclipse to arrive on Tuesday
>
> (*The Guardian*, 7 November 2022)

---

12. Dutch: Special blue supermoon clearly visible in our country.

The preceding Full Moon names relate to the calendar but a *Blood Moon* is an astronomical event. The orbit of the Moon is tilted with respect to the ecliptic (the path of the Sun through the sky) and cuts through this invisible line in two places called the ascending and descending nodes. If a Full Moon takes place at or very near a nodal crossing, then a lunar eclipse occurs. In the case of a total lunar eclipse, the Full Moon passes into the Earth's shadow and is fully immersed in the umbra, the darkest part of that shadow. However, the Moon does not vanish, but usually appears a dark red. This is because the light of the Sun passes through Earth's atmosphere, which filters out the bluer wavelengths, leaving only the faint red light to illuminate the eclipsed Moon. The term *Blood Moon* is not a scientific term but is often used to describe the

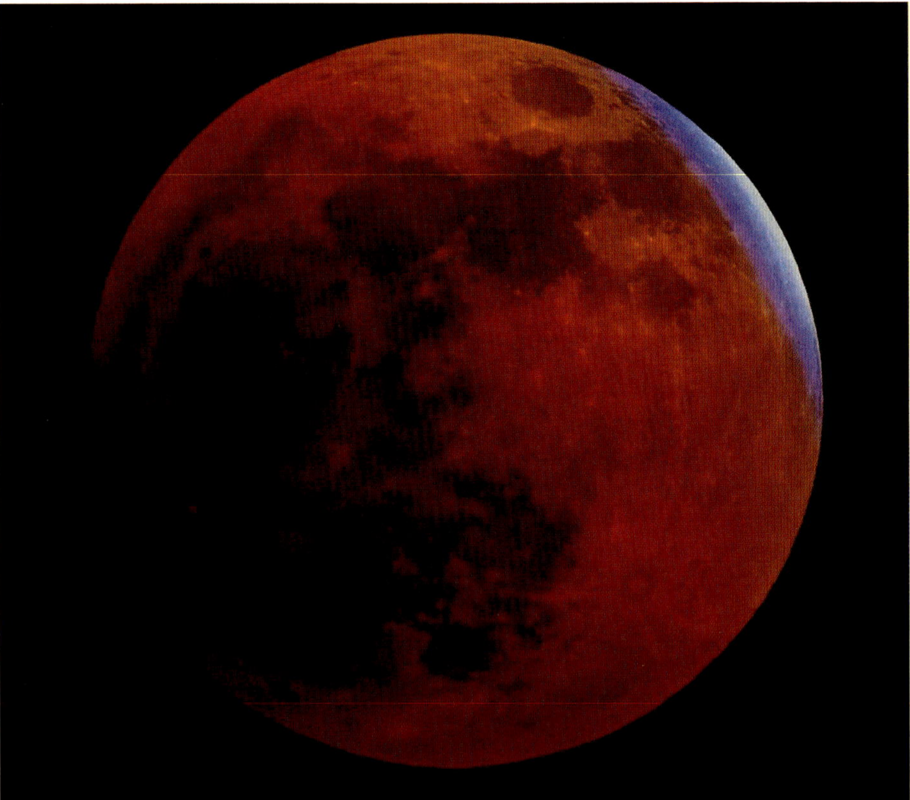

The Full Moon may turn a dark red colour when it moves through the Earth's umbral shadow during a total lunar eclipse. The intensity and darkness of this colour rely on atmospheric conditions on Earth; it is the Sun's rays passing through Earth's atmosphere that illuminate the eclipsed Full Moon. In this photograph from Ciudad de Guatemala, Guatemala, the Blood Moon of 16 May 2022 is not fully immersed in the umbra, leaving a bright sliver of undimmed light along the upper right corner. (Brayan Acajabòn/Pexels)

appearance of a totally eclipsed Full Moon. Even NASA is not averse to using the expression!

Similarly, the terms *Supermoon* and *Micromoon* relate not to the calendar but to the Moon's orbit. The path of the Moon around the Earth is not a perfect circle but rather an ellipse. The point in its orbit nearest to Earth is called perigee and the point furthest is apogee. If a Full or New Moon occurs when it is near perigee, it might be termed a *Supermoon*. Near apogee, it might be called a *Micromoon*. The word *Supermoon* was first coined in the 1970s by an astrologer; it applies to any Full or New Moon occurring when it is within 90% of its perigee distance. Using this definition, there can be three or four Supermoons in a year. A Full Moon near perigee is slightly larger in appearance and slightly brighter than normal. However, the difference between a *Supermoon* and any other Full Moon is not readily discernible to the naked eye. Neither *Supermoon* nor *Micromoon* are well-defined, and like *Blood Moon*, are not scientific terms. The astronomical community prefers the expressions *perigean* and *apogean* full (or new) moons for these phenomena.

The November 2015 Full Moon dominates the night sky over New York City. The Full Moon never fails to excite interest, wonder, and awe, even in modern society. (Reynaldo #brigworkz Brigantty/Pexels)

In Closing

We have seen that Full Moons have many names and many meanings to people around the world. These names reflect the plants and the animals and the traditions of the seasons; they honour spiritual beliefs; they harmonise the lunar calendar with the solar year; and they even reflect real astronomical events. Some names date back centuries or even millennia whilst others are of modern origin. Go out and look at the next Full Moon. What name will you give it? What will it mean to you?

Halley

A Question of Pronunciation

Katrin Raynor and Brian Jones

Edmond Halley

Born in Middlesex on 8 November 1656, Edmond Halley is one of the best-known scientists of all time. He was an astronomer, mathematician and physicist, and held the title of Astronomer Royal from 1720 to 1742. In 1676, Halley travelled to Saint Helena, an island located in the South Atlantic Ocean almost 2,000 kilometres to the west of Africa, to observe and catalogue the stars of the southern sky, a pursuit inspired by John Flamsteed, who was compiling a catalogue of the stars in the northern hemisphere. Whilst on Saint Helena, Halley observed the transit of Mercury across the Sun and later realised that future observations of transits could be used to measure the distance between the Earth and the Sun.

This oil on canvas portrait of Edmond Halley was painted by the Scottish artist Thomas Murray (1663–1734) in around 1690. (Wikimedia Commons/Thomas Murray/Royal Society/Lumos3)

Perhaps Halley's greatest accolade is Halley's Comet, a short-period comet that visits the inner solar system and becomes visible to Earth-based astronomers every 75 years or so, and potentially, therefore, can be seen twice in a person's lifetime. The last appearance of Halley's Comet was in 1986 and it will not appear again until 2061, by which time both writers of this article would be rather advanced in years!

Halley was not the first person to see or record the comet; in fact it can now be traced back as far as 240 BCE when Chinese astronomers first noted its appearance. For centuries the comet was observed and recorded, with perhaps its most famous representation being that which was embroidered onto the Bayeux Tapestry following the comet's return in 1066. It was by using historical observations, together with data from the appearances of comets recorded in 1531, 1607 and 1682, that Halley was able to conclude that the three latter apparitions were actually of the same comet. He subsequently presented his findings, predicting that the comet would reappear in 1758. Unfortunately, Halley was not destined to witness the return, having died at the age of 85 in 1742, although the comet did arrive as predicted, and was subsequently named in Halley's honour.

Halley, Hailey or Hawley?
The pronunciation of Halley's surname has been the cause of debate (and perhaps heated arguments) for years. As you have been reading the opening paragraphs to this chapter, consider how you have been pronouncing his surname. Traditionally, the name has been pronounced to rhyme with 'valley', as suggested in this Royal Astronomical Society drinking song from 1910, the year that the comet returned to the sky:

> Of all the comets in the sky,
> There's none like Comet Halley
> We see it with the naked eye
> And periodically
> The first to see it was not he,
> But still, we call it Halley.
> The notion that it would return
> Was his originally.

This short poem was originally credited to the English astronomer and seismologist Herbert Hall Turner (1861–1930), who had apparently incorrectly written the first line of the song as, 'Of all the meteors in the sky'. However, in the journal *The Observatory*, R.P. Broughton writes that Turner may have not been the true author after all, and its original source remains unknown.[1]

In Brian Harpur's *The Official Halley's Comet Book*, published in 1985, Harpur puts forward a reasoned argument for Halley's surname to be pronounced

---

1. For further information, see 'The Speeches of Spencer Jones', R.P. Broughton, 1984, *The Observatory*, **104**, 273–274 at: **adsabs.harvard.edu/full/1984Obs...104..273B**

'Hawley'.[2] This reason initially comes from the pronunciation of the word 'hall' as 'hawl' giving rise to 'Hawley'. Harpur's research also concluded that during Halley's lifetime, Halley himself wrote his surname both as *Halley* (the actual spelling) and *Hawley* (a phonetic representation), which leads us to the conclusion that Edmond Halley indeed pronounced his own name as 'Hawley'.

During the time Halley was alive, the spelling of his surname varied. It is difficult to nail down the precise pronunciation although there is historical evidence in correspondence to Halley showing

Herbert Hall Turner. (John McCue)

that the misspellings of his surname give a clue in the way that people at the time were pronouncing it. During the period that he lived, people were writing names as they were heard, misspellings including 'Hailey', 'Haley', 'Hayley' 'Hawly' and 'Hawley'; it is significant both that the first three are now generally accepted as being incorrect and that the latter two are the most common.

Using examples from correspondence to Halley from Queen Anne, the Earl of Nottingham and Narcissus Luttrell, Harpur confirmed that 'Hawley' must be the correct pronunciation, with his final decision on this matter attested to naval minutes written by the English diarist and naval administrator Samuel Pepys (1633–1703), who wrote, 'Mr Hawley – May he not be said to have the most, if not to be the first Englishman (and possibly any other) that had so much, or (it might be said) any competent degree (meeting in them) of the science and practice (both) of navigation. And the inferences to be raised therefrom.'

In his book *Brief Historical Relation of State Affairs from September 1678 to April 1714* (Oxford, 1857), the English diarist and historian Narcissus Luttrell (1657–1732) spells Halley's surname in a number of different ways, including Haley and Halley, although one entry tells us that …

<hr>

2. *The Official Halley's Comet Book*, Brian Harpur, 1985, Hodder and Stoughton.

'Captain Hawley, the mathematician, is arrived at Vienna in his way to sound the imperial harbours in the Adriatick Gulph.'

The first two more or less reflect the actual spelling of the surname, although Luttrell's entry quoted here seems to reflect the way that Halley's name was actually pronounced and, along with other sources from those times – such as those from Samuel Pepys – tends to put any argument or controversy regarding the pronunciation of Halley to bed once and for all.

In May 1986, during the 'Halley's Comet – The Comet Returns' episode of the Sky at Night, Patrick Moore raised the question around the pronunciation of Halley, conducting an interview with Harpur to discuss the point, and which can be viewed online.[3] Harpur briefly weighed up the reasons behind his choice of 'Hawley' (as mentioned in his book) and when asked about the return of the comet in 1910, Harpur comically exclaims 'Hallelujah or Hawllelujah!'.

Oil on canvas portrait of Samuel Pepys, dating from 1666, by the English portrait painter John Hayls (1600–1679). (John Hayls/National Portrait Gallery/NPG 211)

Brian Jones, co-writer of this article and co-editor of the book you are holding, is a staunch advocate of 'Hawley' and will often correct anyone who pronounces it any other way. Drawing on contemporary sources that Harpur also drew upon, in 2014 he (Jones) published a short article about Edmond Halley on his website Starlight Nights writing, 'From this (Pepys') diary entry, and other phonetic variations of the period, it is evident that the name was pronounced 'Hawley', with the 'Hall' part of the surname rhyming with 'ball'. This is a reflection of the pronunciation that Halley himself preferred and which, out of respect to him, the one that we should certainly use today.'[4]

Many people nowadays say 'Hailey', a legacy of the rock and roll group Bill Haley and His Comets which formed during the late-1940s under the name

3. The Sky at Night, May 1986, The Sky At Night – 'The Comet Returns' – Halley's Comet - BBCTV - 1985 **youtube.com/watch?v=am2DP70kgs8**

4. *Edmond Halley*, Brian Jones, 15 January 2014, Starlight Nights **starlight-nights.co.uk/edmond-halley**

This image, taken from 'Round Up of Rhythm', a musical short film released in 1954 by Universal International and produced and directed by Scottish-born Will Cowan (1911–1994), shows Bill Haley and His Comets during what was their motion picture debut. The group was originally formed in around 1949, performing under the name of Bill Haley and the Saddlemen until 1952. It was a friend of Bill Haley who at this time suggested that Haley rename his band Bill Haley and His Comets, citing the common mispronunciation of the name of the famous Halley's Comet to rhyme with Bailey. The change of name to incorporate 'His Comets' rather than 'the Saddlemen' in 1952 may well have laid the seeds for the ongoing mispronunciation of the famous comet. (Wikimedia Commons/Universal International/Klau Klettner/HydraRecords/Bill Haley Museum, Munich, Germany)

of Bill Haley and the Saddlemen. In 1952 it was suggested that the band change their name, replacing 'the Saddlemen' with 'His Comets', a change which was inspired by the supposed pronunciation of Haley.

At the last return of the comet in 1985, the astronomy writer, broadcaster and lecturer Ian Ridpath telephoned representatives of 36 Halley families and asked them how they pronounced their own name. His results found that two pronounced it 'Hailey', and two others said 'Hawley'. The rest, 89 per cent of Ian's sample, used the traditional pronunciation, which is also commonly used by astronomers engaged in comet study. Ian concluded that the traditional pronunciation clearly has the support of the vast majority of 'present-day Halleys', although the most important consideration for astronomers when pronouncing the name may be to do so in the way that Halley himself did.

Inspired by Ridapth's telephone poll, which was undertaken before the days of social media, Katrin Raynor – the other co-writer of this article and co-editor of this book – took to X (Twitter) and Facebook to conduct a similar poll asking how people pronounce Halley. This time however, instead of targeting people with the name Halley, she was interested to ask her followers on X and the Comet Watch Facebook group for their opinion. On X, the results concluded that 13 people who engaged in the poll pronounced it 'Halley', six people say 'Hailey' and one person says 'Hawley'.

The Facebook poll provided a set of more interesting results and comments to consider. Out of 89 votes, 63 voted for 'Halley', 10 voted for 'Hailey' and 12 voted for 'Hawley'. One member of the Facebook group suggested 'HaLee' but it is unclear how this differs from 'Halley' (to rhyme with valley). Interestingly, four people voted for 'HaLee'. One member commented that, while despite himself saying 'Hawley' and to avoid being seen as pretentious and for people to understand the topic of conversation, he would say 'Halley' (to rhyme with valley), which perhaps can be used to explain why radio and television presenters may forsake the actual written evidence (and, some would say, the evidential written word of the man himself) and use that same pronunciation.

Unfortunately, there are no known descendants of Halley to help draw this debate to a close and it is clear that his name is still, and will in the future, be pronounced in different ways, although after reading the compelling evidence, the writers of this article will certainly be respecting the great astronomer's legacy and pronouncing the surname 'Hawley'.

Uranus

What's in a Name?

Brian Jones and Katrin Raynor

Discovered in 1781 by astronomer William Herschel (1738–1822), Uranus is the seventh planet out from the Sun. An extreme axial tilt of almost 98 degrees means that for around a quarter of its year (21 Earth years), the Sun shines directly over each pole, whilst the other half of the planet is in darkness. Another factor that makes this planet unusual is that it spins in the opposite direction to all the other planets, with the exception of Venus which also spins 'backwards'.

Although Uranus is indeed something of a unique planet, it holds yet another distinction; of all the planets in the Solar System, Uranus has raised by far the most controversy and misunderstanding over the pronunciation (or mispronunciation) of its name …

In the wake of its discovery, the task of coming up with a name for this hitherto unknown world fell on the shoulders of its discoverer. When asked to name the planet, Herschel settled upon *Georgium Sidus* (George's Star) or the 'Georgian Planet', in dedication to King George III who was the British monarch at the time of the planet's discovery. Understandably, this name was unpopular, so other proposals were put forward including *Herschel*, a name proposed by French astronomer Joseph Jérôme Lefrançois de Lalande (1732–1807), and even *Neptune* (which itself wasn't discovered until 1846 and was thus the next planet to be named).

In March 1782 the German astronomer Johann Elert Bode (1747–1826) proposed

This image of Uranus was captured by the *Voyager 2* probe on 23 January 1986. The pronunciation of the name of the seventh planet out from the Sun has often formed (perhaps appropriately) the butt (pun intended) of many jokes. (NASA/Voyager 2/Ardenau4)

naming the new planet *Uranus*, his argument being based on the fact that Saturn was the father of Jupiter, and Herschel's new discovery should therefore be named after the father of Saturn. Interestingly, in Great Britain, Uranus was not officially accepted as the planet's name until 1850 when, up until that time, the planet had been called *Georgium Sidus*.

Among all the astronomical phenomena, Uranus stands virtually alone in the amusement stakes for those who continue to stifle a guffaw when the planet's name is mentioned. Science lessons will still see a proportion of students await the teacher's use of the word in order to trigger a reaction from both the person stating 'Uranus' and the associated response from their classmates.

The astronomer Johann Elert Bode can hardly have imagined the controversy to come following his choice of name for the newly-discovered planet. (John McCue)

A growing number of probes have and continue to visit the planets in our solar system, each subsequent probe armed with a more elaborate and sophisticated arsenal of instruments in the hope of examining and ultimately discovering more about our planetary neighbours. Some probes would visit multiple planets, like *Voyager 2*, which after visiting Jupiter and Saturn, made a fly-by of Uranus before reaching out to Neptune. Other probes, like Russia's numerous *Venera* probes, would have just one destination, Venus. In the case of *Venera*, and indeed the link with fertility, one may associate corresponding pictures as perhaps being more of a gynaecological nature, rather than the hostile, barren alien surface that was presented.

In English-language popular culture, humour is often derived from the common pronunciation of Uranus's name, which resembles that of the phrase 'your anus'. Indeed, as the writer Daniel Craig, who worked for Philadelphia-based media publisher PhillyVoice between 2015 and 2018, informs us in his entry for 20 June 2017 headed 'Very nice job with these Uranus headlines, everyone'[1] that 'NASA wants to probe Uranus in search of gas' (*BGR*); 'NASA wants to probe deeper into Uranus than ever before' (*Metro*); or even 'Uranus Smells Like Farts' (*Gizmodo*)

Now, consider how you have been pronouncing Uranus when you have been reading the above introduction. Do you say yoor-uh-nus or yoo-rey-nus? These pronunciations are taken from **dictionary.com** and seem to be the two

---

1. See: **phillyvoice.com/very-nice-job-these-uranus-headlines-everyone**

widely accepted pronunciations in the modern-day English language. Some people may snigger at the latter because it sounds rude, and for the perhaps overly-prudish out there, it is a word that just cannot be said! There are many jokes involving the pronunciation although, humour aside, the pronunciation of Uranus is interesting, and it seems, even now, there is no agreed way to say the planet's name.

Historically, the name Uranus or Ūranus comes from the Latinised version of the Greek god of the sky *Ouranós*, the personification of Heaven and ruler of the world, which is pronounced oo-rah-noss. Both the Greek and Latin pronunciations do sound similar with the 'a' in Uranus sounding like the 'a' in apple. Of course, different languages across the globe pronounce letters and words differently and it is surely for this reason that over the centuries we have ended up with both the yoor-uh-nus or yoo-rey-nus pronunciations.

A few searches online provide some interesting videos around the pronunciation of the planet's name. According to American planetary scientist Kevin Grazier, who on a YouTube video called 'Scientists Explain How Do You Pronounce URANUS', he explains that the pronunciation of Uranus was changed for media purposes when the *Voyager 2* spacecraft encountered the planet in 1986. To avoid any laughing and possible embarrassment on the part of the public and media, the 'yoor-uh-nus' pronunciation was used when announcing the success of the *Voyager 2* program. Grazier himself said in a further interview in 2016 with Fraser Cain that the planet should be pronounced 'yoor-AH-nus'.

The editor, designer and writer James Harbeck provides a useful insight into the pronunciation of the planet in his video Pronunciation tip: Uranus.[2] He briefly talks about the historical pronunciation and mentions that Uranus is perhaps pronounced 'yoo-rey-nus' simply because that is how the word looks – Ur-anus. He also makes an interesting point around the pronunciation of uranium, a radioactive substance used in nuclear power generation and radiation shielding. We do not say yoo-rah-nium but instead yoo-RAY-nium.

We can also consider the pronunciation of Uraniborg, an observatory and laboratory built by Danish astronomer Tycho Brahe (1546–1601) on the island of Ven which was part of Denmark at the time. The name Uraniborg was chosen for the observatory as a dedication to Urania, the Muse of Astronomy. The Greek name for Urania is *Ouranía* which according to Wikipedia, is pronounced yoor-AY-nee-a. Again, it seems that listening from a few clips online, Uraniborg is pronounced U-RAY-niborg or U-RAH-niborg.

There appears to be no right or wrong way to say Uranus – we all know what is being meant when speaking about the planet, no matter how it is said. In conclusion, there is a choice of method for pronouncing Uranus:

---

2. Pronunciation tip: Uranus: **youtube.com/watch?v=pJbMJzkczgM**

Oo-rah-noss
Yoor-uh-nus
Yoor-ah-nus
Yoo-rey-nus
… and, let's not be coy … Your-anus

One thing is clear however, that in this modern day and age, astronomers and Joe Public do not use *Ouranós* but given the pronunciation of the Greek word, some would (perhaps erroneously) suggest that the correct way of saying the planet's name is yoor-uh-nus or yoor-ah-nus.

To sum up, various dictionary sources inform us that the name of the planet can be pronounced in either of the two popular/common ways. It does seem, however, that prior to the *Voyager 2* mission, the overwhelming majority of

Please, Sir . . . can you tell us more about the methane in Uranus . . ?

This cartoon epitomizes just one of the many amusing circumstances surrounding the name of the planet Uranus and its (correct) pronunciation. However, it is not only classroom-based pupil-teacher feedback that causes amusement; indeed, the pronunciation must also have created merriment in many podcasts, lecture theatres, and television and radio studios – and with their doubtless-appreciative viewing audiences – over the years (or perhaps should have done, had the individuals concerned had not shied away from 'going with the flow' and making the most of this famous idiosyncrasy of the English language). (John McCue)

people pronounced the name of the planet as Your-anus. Yet since the advent of the somewhat overzealous need of modern day broadcasters and so on not to 'offend' (as if) the supposedly-delicate ears of the public, the slightly more 'sensitive' methods of pronunciation seem to have taken precedence. This is perhaps unfortunate, and we strongly suggest that such broadcasters really do need to lighten up rather than worrying about things quite so much!

And perhaps the item that says it all …

The journalist and broadcaster Alastair Burnet introduced Jupiter's recently discovered moon 'Bumhole' (c1985)[3] on the sometimes hard-hitting but always excellent British satirical puppet show Spitting Image. Alastair's candid and highly-entertaining delivery certainly emphasises the silliness of it all and underlines the fact that we really should all enjoy the idiosyncrasies of the English language rather than kowtowing to what certain areas of the media regard as being the 'appropriate thing' to do …

Perhaps Uranus itself wishes it could be named differently but then again, why should it be called anything else? The name sits nicely as a cosmic milestone as we work our way outwards from Mercury in the solar system. Perhaps some comfort is to be found from neighbouring planets and their moons. You can just hear one of Jupiter's (other) moons making an apology to Uranus on behalf of us – often over-cautious – simpletons here on planet Earth: "Io you an apology on behalf of humankind."

The ideal and (many would agree) conclusive, way of verifying the correct pronunciation would be to hear a recording made prior to it being 'interfered with' by the likes of NASA and the BBC. Such irrefutable evidence is indeed available, courtesy of the noted English stage and film actor Cedric Hardwicke (1893–1964) who was the post-titles narrator for the 1953 film The War of the Worlds. In the narration at the beginning of the film he describes the other planets as potential abodes for intelligent life. During this sequence he clearly pronounces the name of the planet Uranus as 'Your-anus', which plainly puts the argument to rest. After all, if an accomplished narrator like Cedric Hardwicke pronounces it that way, then that's the way it is!

The writers of this article will, of course, leave the pronunciation of the name up to you, but we will continue to pronounce the planet as Your-anus. It really does come down to a matter of preference and whether or not you can hold in a giggle! So, now you have read this chapter, go out and enjoy your newfound freedom of expression, albeit in the knowledge that you may not be considered for a role on certain astronomy-themed television programmes!

---

3. Alastair Burnet on Spitting Image talking about Jupiter's recently discovered moon 'Bumhole' is available to view at: **youtube.com/watch?v=pHp9Cakv2Fg**

Astronomy Enters Popular Culture

Katrin Raynor and Brian Jones

In this modern day of technology, social media, e-books and websites, astronomy really is just one click away at any time of day or night. It wasn't that many years ago that the first Moon landing was transmitted into millions of households, families crowding around small black and white televisions to catch the first glimpse of humans landing on the Moon. Now, consider how much has changed in that time. We can access telescopes from the other side of the world from the comfort of our own home, or watch the crew of the *International Space Station* zip around the Earth as NASA TV transmits live footage from the station. Cheap flights enable millions of us to jet across the other side of the world to take in a total solar eclipse, or we can watch the dancing greens and reds of aurorae in colder climes from the decks of cruise ships.

Eclipse-watchers in Chile, South America
Eclipse-chasing seems to have become something of a global cultural event. Here we see a group of eclipse chasers from around the world gathered in Chile, South America to watch the total solar eclipse which took place on 3 November 1994. Although the eclipse itself lasted only a little over four minutes, many of the observers seen here will no doubt have gathered yet again to watch the next similar event. The image seen here is a composite of photographs of astronomers watching the eclipse and a painting of the eclipse itself. (© David A. Hardy/astroart.org)

Awaiting the Eclipse
Seasoned eclipse chasers Neil Haggath and Peter Strugnell await the total eclipse of 8 April 2024 at Valley Mills, Texas. The phrase 'eclipse chaser' is perhaps something of a misnomer. After all, it isn't the observers who do the chasing, but the direct opposite. The so-called eclipse chasers gather together at a point *ahead* of the eclipse and wait for the event to reach *them* … hopefully under a cloudless blue sky! (Neil Haggath/ Peter Strugnell)

Finding the Pole Star
One of the most widely-known astronomical facts amongst school children, scouting groups and organisations and even the public at large, is that the star Polaris, commonly referred to as the 'North Star' or the 'Pole Star' in the constellation of Ursa Minor (the Little Bear), lies nearly in a direct line with the Earth's rotational axis, a seemingly fixed point located directly above the Earth's north pole. Contrary to what the majority of people believe, Ursa Minor is by no means the most prominent constellation in the sky, although it stands out reasonably well due to the fact that the area around it is devoid of bright stars. Ursa Minor is easily located by using the two end stars in the 'bowl' of the Plough – formed from seven of the brightest stars of Ursa Major – as

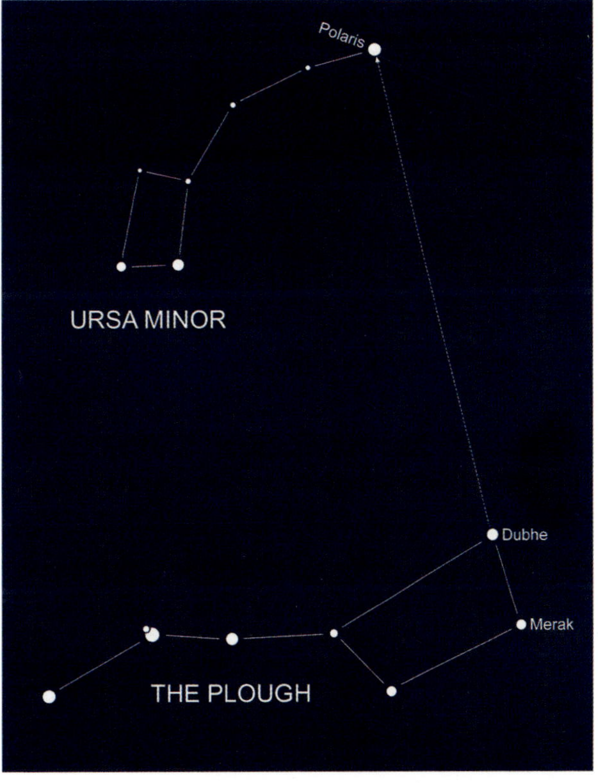

pointers, a line from Merak through Dubhe eventually leading you to Polaris. A number of myths and legends surround Ursa Minor, one of the most popular being that the constellation represents Arcas, the son of Zeus and Callisto. Arcas was turned into a bear along with his mother (represented by Ursa Major), after which Zeus placed both bears next to each other in the sky. (Brian Jones/Garfield Blackmore)

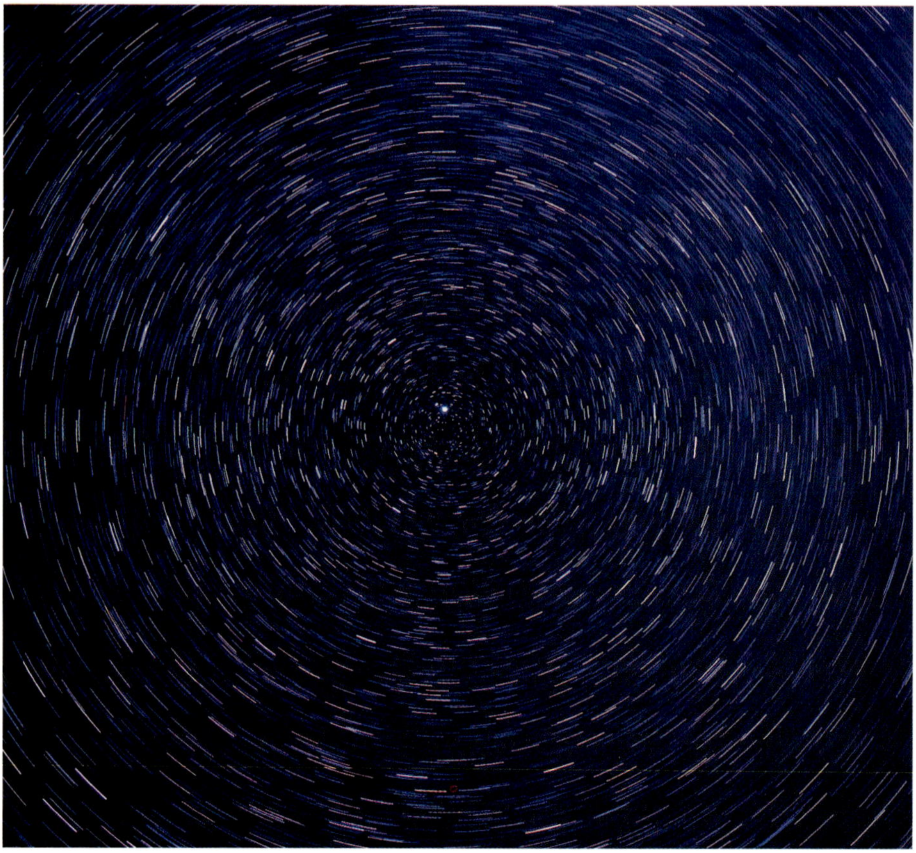

Polaris lies nearly in a direct line with the Earth's rotational axis – a point in the sky known as the north celestial pole – the effect of this being that the Earth's rotation causes the stars and constellations around the celestial pole to appear to travel around it in concentric circular paths every 24 hours. Long exposure photographs centred on the celestial pole will produce circular star trails, as seen here, which clearly reflect the rotation of the Earth around its axis. Polaris marks the position of the north celestial pole, its southern counterpart being located in the tiny constellation Octans (the Octant). Although observers in the southern hemisphere have no particularly bright star marking the position of the south celestial pole, long exposure images will still produce the same effect as seen here. (Fred Moon/Unsplash Licence)

A Van Gogh inspired sky hangs over members of the York Astronomical Society in this acrylic painting by Marion Tasker depicting a public star party. (Marion Tasker/**maidbymarion.co.uk/** Martin Dawson)

Sidewalk Astronomy

An increasing number of astronomical societies host public observing sessions in order to engage with the community and to inspire them with the wonders of the night sky. Here we see members of Sussex Sidewalk Astronomers, gathered together along with an impressive array of telescopes, their members actively seeking to encourage and nurture those who wish to learn more about the wonders of astronomy. As members of Sussex Sidewalk Astronomers point out, 'The universe doesn't charge us to look at it, so why should we?' (Chris Woodcock)

Star Parties

Star Parties are an entertaining and inexpensive way to engage the local community in astronomical events. This image shows York Astronomical Society playing host to members of the public during one of their frequent star parties. During these events, non-astronomers are treated to views of a wide range of celestial objects, ranging from the Moon and planetary members of the Solar System out to a whole host of wonderful sights located well beyond our region of space including star clusters, nebulae and galaxies. For many, these will be their first-ever glimpse of the universe with optical aid, and without a doubt, the first step on a tall (and potentially expensive!) astronomical ladder. (Martin Dawson)

Meteor Watching
Members of the Guildford Astronomical Society gather for a meteor observing event on the evening of 7 January 2025, their target being the annual Quadrantid meteor shower. Meteor watching is becoming increasingly popular, not only with astronomical societies but also with the public, mention of current meteor showers often featuring on television weather forecasts. Readers who are familiar with the night sky will no doubt recognise the brilliant Sirius – the leading star in the constellation of Canis Major (the Great Dog) and the brightest star in the sky – in the background, with the faint but distinctive shape of Lepus (the Hare) to its right. (Rachel Dutton/Guildford Astronomical Society)

On a much smaller scale, affordable telescopes, astro-tourism and local astronomical societies have introduced astronomy into local communities, some thanks to the help of media outlets such as news channels who report on notable upcoming events, piquing people's interest. Astronomy outreach has become an important part of what seems like everyday life. From organizing a meteor watch event or attending an astronomical society's star party, even standing on a pavement on your way to work to witness a solar eclipse, the wonders of space really are closer to home than you may think.

Watching the Aurora
The edge of a nearby forest provides a scenic background to this image showing a group of aurora watchers gazing up at an impressive display of the northern lights. Aurorae are mainly observed in the high-latitude regions of the Arctic and Antarctic, these events being known as the aurora borealis (northern lights) and aurora australis (southern lights). Trips to see aurorae, either by sea, rail or air, are becoming a common event, with many travel companies offering tour packages which include opportunities for passengers to see for themselves these often-stunning natural light displays which occur many miles above the Earth's surface. (Adobe Stock/Parilov)

Dr Jenny Shipway

Science Communication | Education

Dr Jenny Shipway is a leading figure in astronomy education and public engagement, based in the UK and working worldwide.

With over 20 years of experience, she's known for making complex scientific ideas engaging and accessible for people of all ages. She has held education roles at every level in science centres and planetariums, and enjoys management, coordination, and delivery of large-scale collaborative programmes. She recently coordinated UK National Astronomy Week and leads the 100 Hours Under One Sky project.

Jenny enjoys creating resources for diverse audiences. She writes for the web and has written and consulted for nearly twenty international planetarium films.

Having a strong interest in education research, Jenny also offers talks and training for science communicators about how people learn, with a focus on practical ways to improve their practice. She has published articles on this topic in professional journals.

If you're involved in project planning or development, Jenny can help you make a bigger impact. Whether through consultation, training, or partnerships, she brings a deep knowledge of both science and effective communication to the table.

You can learn more or get in touch via her website at **jennyshipway.com**

- IAU Office of Astronomy Outreach, National Outreach Coordinator (UK)
- International Planetarium Society, Education Committee Co-Chair
- International Network of Amateur Astronomy Organisations, founder and Co-Chair
- Federation of Astronomical Societies (UK), International Liaison
- AstroBoost (UK, supporting astronomy society outreach), founder and Director
- AstroMailbox (UK astronomy education email list), founder and Coordinator
- British Association of Planetaria, Lore-Keeper and former President

Astronomy Outreach

Astronomy is for Everyone

Jenny Shipway

Astronomy has always been part of our human experience, with evidence of people taking interest in the sky since pre-history. There are countless cultural myths and legends relating to the incredible view we enjoy from our small planet; stories told over millennia as people looked up into the darkness.

More recently, artificial lighting, expanding cities and busy lives have made it easy for people to miss this inspiring and deeply human experience. But at the same time we are pushing forward our understanding of the universe, and looking toward the possibility of visiting other planets. Our societies are increasingly at risk of adverse space weather and we are deeply in need of viewing our planet as a whole. For these reasons, astronomy outreach is more important than ever.

Happily, there is a huge range of organisations and individuals across the world passionate about taking astronomy into schools and to different public audiences.

Perhaps the largest network, the International Astronomical Union (IAU)'s Office for Astronomy Outreach (OAO)[1] has National Outreach Coordinators in over 120 countries supporting and carrying out activities. As well as running talks and events for different audiences, coordinators have built astronomy libraries in Haiti and Benin, delivered workshops for teachers in Algerian refugee camps, and run an international project following the path of a recent eclipse across Latin America.

Most countries also have activities run by universities, observatories, planetariums, and science and discovery centres. Many also have clubs/societies, and businesses offering education and entertainment events.

Amateur astronomers play a vital role in this ecosystem. They are spread widely, including in regions that are less-often visited by universities and science organisations. And, importantly, they share their love of astronomy

---

1. IAU Office for Astronomy Outreach: **iauoutreach.org/es**

within their own local communities. We know that people need to see 'people like them' involved in science if they are to feel like it's something that might be 'for them'.[2]

Many astronomers can remember a special moment when their curiosity was sparked and they discovered their interest. Perhaps when looking through a telescope for the first time. But we should understand that such interest doesn't emerge from a void – the power of the spark is to crystallise a wide variety of accumulated prior thoughts and knowledge.

And so a truly astronomy-engaged society should also have astronomy embedded within everyday life, providing a broad cultural base where astronomy is regularly encountered and well-regarded. This means storybooks and TV shows and everyday knowledge about the stars. This is the foundation from which people can reach those special moments of inspiration and be supported by those around them to pursue their interests.

We need to reach everyone. Astronomy communicators must travel to regions where there are no local providers, including to geographically isolated rural areas in developing world countries (and they do!). Meanwhile, fixed and mobile planetariums can bring the beauty of a dark, rural night sky into large cities.

To make sure everyone is welcome and included, activities must be diverse, with options that are culturally, physically and emotionally accessible to different groups. To go beyond majority audiences, activities must be appropriately designed, working with the group(s) in question to make sure that the activity is interesting and relevant. This also builds trust, which is important for groups who have far too often seen events aimed *at* them that were not *for* them.

Individual astronomers or smaller organisations may not always have the time to develop accessible activities from scratch. But by linking up organisations both nationally and internationally, it is possible to share skills, ideas, and resources. Shared resources can include lesson plans, hands-on activity ideas, information/images, and even instructions for building telescopes or other equipment using locally-available resources.

This sharing has been powerful in improving inclusion of under-served audiences, especially the visually impaired. Accessible, hands-on activities include 3D printed star and galaxy maps, available from projects including Tactile Universe (based in the UK and led by a blind astronomer).[3] Projects like

---

2. Macdonald, A., (2014). "'Not for people like me?' Under-represented groups in science, technology and engineering," (WISE Campaign, Leeds, UK)
3. Tactile Universe: **tactileuniverse.org**

Stelle per Tutti (Italy)[4] offer a variety of accessibility resources to astronomy communicators.

Beyond activity resources, organisations including the IAU OAO and many planetarium associations offer training and networking opportunities, where astronomy communicators can gain skills in project management and hear about each others' projects' successes and failures.

International networks also lead to collaborative events, linking people across borders and reminding us that we all live on the same planet. Examples include: the 100 Hours of Astronomy IAU OAO global event that encourages people to carry out astronomy activities during a 100-hour period; Astronomers Without Borders' redistribution programme for solar-eclipse glasses;[5] and the A Arte da Astronomia virtual gallery[6] that exhibits works from children in Angola, Brazil, Cape Verde, Mozambique and Portugal.

A rich and diverse ecosystem of communicators and activities is vital if we want to move towards a world where astronomy is embedded in cultures, where everyone feels a personal connection to our cosmos, benefits from an understanding of their place in the universe, and has an appreciation for how space affects our daily lives.

---

4. Stelle Per Tutti: **uai.it/stellepertutti**
5. Astronomers without Borders: **astronomerswithoutborders.org**
6. A Arte da Astronomia: **iaunocbrasil4.wixsite.com/a-arte-da-astronomia**

Our Contributors

Stuart Atkinson is a writer and amateur astronomer living in Kendal (as in the Mint Cake, yes…) in Westmorland in the Lake District National Park. He has been writing children's astronomy books since 1988, and has now had 14 published. He is best known for *A Cat's Guide to the Night Sky*, published by Laurence King in October 2018. Since publication the book has been translated into more than 20 languages worldwide. For the past 25 years he has also been an astronomy and "space" editor/consultant for numerous other publishers. When Stuart is not writing or editing astronomy books he is kept busy writing features and articles about astronomy for a number of monthly astronomy magazines, including the *Sky at Night* magazine, *All About Space* and *Astronomy*. He is very active on Facebook and has more than 10,000 X followers, who enjoy looking at the images he takes of the night sky and the NASA images of Mars that he processes. One of his images of the Martian landscape was viewed over half a million times on X.

Neil Haggath has a degree in astrophysics from Leeds University and has been a Fellow of the Royal Astronomical Society since 1993. A member of Cleveland and Darlington Astronomical Society since 1981, he has served on its committee since 1989. Neil is an avid umbraphile, clocking up six total eclipse expeditions so far to locations as far flung as Australia and Hawai'i. Four of them were successful, the most recent being in Jackson, Wyoming on 21 August 2017. In 2012, he may have set a somewhat unenviable record among British astronomers - for the greatest distance travelled (6,000 miles to Thailand) to NOT see the transit of Venus. He saw nothing on the day … and got very wet!

David A. Hardy is the longest-established space artist, illustrating his first book, for Patrick Moore, in 1954 and continuing to work with him on books and for 'The Sky at Night'. David is European Vice-President of the International Association of Astronomical Artists (IAAA) whose website is at **iaaa.org** and recipient in 2001 of the Rudaux Memorial Award for services to astronomical art. In March 2003 he became one of only a handful of artists to have had an asteroid named after him. David has also written or collaborated on many non-fiction books and has a published novel *Aurora: A Child of Two Worlds*. David Hardy's work, including originals, prints and DVDs, is available from AstroArt, 99 Southam Road, Hall Green, Birmingham B28 0AB (tel. 0121 777 1802), or go to his website **astroart.org**

David M. Harland gained his BSc in astronomy in 1977 and his PhD in computer science in 1981. He has lectured in computer science, worked in industry and managed academic research. In 1995 he 'retired' in order to write on space themes.

David Harper, FRAS has had a varied career which includes teaching mathematics, astronomy and computing at Queen Mary University of London, astronomical software development at the Royal Greenwich Observatory, bioinformatics support at the Wellcome Trust Sanger Institute, and a research interest in the dynamics of planetary satellites, which began during his PhD at Liverpool University in the 1980s and continues in an occasional collaboration with colleagues in China. He is married to fellow contributor Lynne Marie Stockman.

Brian Jones hails from Bradford in the West Riding of Yorkshire and was a founder member of the Bradford Astronomical Society. He developed a fascination with the night sky at the age of five when he first saw the stars through a pair of binoculars, his interest in astronomy eventually taking him into the realms of writing sky guides for local newspapers, appearing on local radio and television, teaching astronomy and space in schools and, in 1985, to becoming a full time astronomy and space writer. As well as being Editor of the *Yearbook of Astronomy* since the 2017 edition, Brian has penned around 20 books to date, which have covered a wide range of astronomy- and space-related topics for both children and adults, and his journalistic work includes writing articles and book reviews for several astronomy magazines as well as for many general interest magazines, newspapers and periodicals. His passion for bringing an appreciation of the universe to his readers is reflected in his writing. The minor planet 45689 Brianjones is named after him.

John McCue graduated in astronomy from the University of St Andrews and began teaching. He gained a PhD from Teesside University studying the unusual rotation of Venus. In 1979 he and his colleague John Nichol founded the Cleveland and Darlington Astronomical Society, which then worked in partnership with the local authority to build the Wynyard Planetarium and

Observatory in Stockton-on-Tees. John is currently double star advisor for the British Astronomical Association.

Mary McIntyre is an amateur astronomer and astronomy communicator based in Oxfordshire, England. She is a keen astrophotographer but also loves creating and teaching astronomy sketching and art. She delivers astronomy related presentations to astronomy societies, camera clubs, history clubs, U3A groups, the Women's Institute and other local community groups. She also delivers astronomy talks and runs astronomy sketching and craft workshops for children in schools, Beavers/Cubs/Scouts and children's astronomy clubs. She is passionate about astronomy outreach and was awarded the 2021 Sir Patrick Moore Prize by the British Astronomical Association for her outreach activities. She is a Fellow of the Royal Astronomical Society and a regular contributor to *Sky at Night* magazine and the *Yearbook of Astronomy*. She has made numerous radio appearances, is a co-presenter of the Comet Watch radio show and a regular panel member on the Space Oddities live panel show. She also runs the UK Women in Astronomy Network which helps to provide inspiration to women and girls who want to pursue a career in astronomy.

Dr Robert Massey is currently the Deputy Executive Director of the Royal Astronomical Society and Visiting Honorary Professor at the University of Sussex. He began his career in astronomy at the Universities of Leicester and Manchester, completed his PhD studying the Orion Nebula, worked as a teacher in Brighton, and spent eight years at the Royal Observatory Greenwich. These days he looks after the outward facing work of the Royal Astronomical Society, and enjoys the darker night skies of his new family home in Sussex.

Jonathan Powell worked at BBC Radio Wales as their astronomy correspondent and is currently astronomy and space correspondent for *The National* (an online newspaper for Wales) and a columnist at the *South Wales Argus*. He is also a contributor to CAPCOM, an online magazine which promotes astronomy and spaceflight to the general public, in addition to which he has presented on commercial radio at Sunshine FM in Worcester, Brunel FM in Swindon, Bath FM in Bath, and on the astronomy and space dedicated radio station Astro Radio UK. As well as writing several articles for the *Yearbook of Astronomy*, Jonathan has penned three books on astronomy – *Cosmic Debris: What It Is and What We Can Do About It*; *Rare Astronomical Sights and Sounds* (which was selected by *Choice* magazine as an Outstanding Academic Title for 2019); and *From Cave Art to Hubble: A History of Astronomical Record Keeping*.

Katrin Raynor lives in Pontypridd, Glamorgan, South Wales and is a Fellow of the Royal Astronomical Society and Royal Geographical Society. She is an amateur astronomer and philatelist, and in her spare time writes articles and interviews for popular astronomy magazines including *Sky at Night* magazine and the Society for Popular Astronomy magazine, *Popular Astronomy*. Katrin is also a co-presenter on the Sky at Night's podcast, Star Diary. The minor planet 446500 Katrinraynor is named after her.

Peter Rea has had a keen interest in lunar and planetary exploration since the early 1960s and frequently lectures on the subject. He helped found the Cleethorpes and District Astronomical Society in 1969. In April of 1972 he was at the Kennedy Space Centre in Florida to see the launch of Apollo 16 to the moon and in October 1997 was at the southern end of Cape Canaveral to see the launch of Cassini to Saturn. He would still like to see a total solar eclipse as the expedition he was on to see the 1973 eclipse in Mali had vehicle trouble and the meteorologists decided he was not going to see the 1999 eclipse from Devon. He lives in Lincolnshire with his wife Anne and has a daughter who resides in Melbourne, Australia.

Marc Read was born on the island of Jersey, and now lives in Newcastle upon Tyne with his family. He studied the history and philosophy of science at Oxford, Cambridge, and London. His research challenged several traditional accounts of how pre-telescopic astronomers viewed and used philosophy; he concentrated on how astronomical and astrological theories influenced, and were influenced by, prevailing religious and cultural movements, this being the subject of his first book, *New Stars for Old*. He now teaches philosophy, world religions, and some physics, and spends his free time with classic films and overly-complicated board games.

Jenny Shipway is a professional science communicator with a background in science centres and planetariums, and now works as a consultant, writer, and project manager. She coordinated National Astronomy Week 2025 and is involved with a number of national and international networks. She has a particular interest in education, and in bringing different people and organisations together to share skills and resources to achieve more than would otherwise be possible.

Lynne Marie Stockman holds degrees in mathematics from Whitman College, the University of Washington and the University of London. She has studied astronomy at both undergraduate and postgraduate levels, and is a member of the Astronomical Society of the Pacific. A native of North Idaho, Lynne has lived in Britain since 1992. She was an early pioneer of the World Wide Web: with her husband and fellow Yearbook contributor David Harper, she created the web site **obliquity.com** in 1998 to share their interest in astronomy, computing, family history and cats.

Philip Wallace has been interested (his family would say obsessed) with astronomy, space and spaceflight from a very young age, reading every book he was able to lay his hands on – even when he didn't understand them – and watching every documentary he could. This passion was fuelled not just by the reality of the cosmos, but by our dreams of it as well, with just as much time reading science fiction as textbooks. He decided to study astronomy at university, graduating in 2013. Since then he continues it both as a hobby and a passion, joining the Cardiff Astronomical Society in 2015 and being elected its Chairman in 2016. Philip has written articles and given talks to anyone that asks, including the general public, other astronomical societies or his colleagues at work. In 2018 he was elected a Fellow of the Royal Astronomical Society, his astronomical journey continuing to this day and hopefully far into the future.